T0372253

School Culture Improvement

RIVER PUBLISHERS SERIES IN INNOVATION AND CHANGE IN EDUCATION - CROSS-CULTURAL PERSPECTIVE

Volume 7

Series Editor

XIANGYUN DU
Aalborg University
Denmark

Editorial Board

- Alex Stojcevski, Faculty of Engineering, Deakin University, Australia
- Baocun Liu, Faculty of Education, Beijing Normal University, China
- Baozhi Sun, North China Medical Education and Development Center, China
- BinglinZhong, Chinese Association of Education Research, China
- Bo Qu, Research center for Medical Education, China Medical Education, China
- Danping Wang, The Department of General Education, Technological and Higher Education Institute of Hong Kong
- Fred Dervin, Department of Teacher Education, Helsinki University, Finland
- Kai Yu, Faculty of Education, Beijing Normal University, China
- Jiannong Shi, Institute of Psychology, China Academy of Sciences, China
- Juny Montoya, Faculty of Education, University of ANDES, Colombia
- Mads Jakob Kirkebæk, Department of Learning and Philosophy, Aalborg University, Denmark
- Tomas Benz, Hochschule Heilbronn, Germany

Nowadays, educational institutions are being challenged when professional competences and expertise become progressively more complex. This is mainly because problems are more technology-bounded, unstable and ill-defined with the involvement of various integrated issues. To solve these problems, it requires interdisciplinary knowledge, collaboration skills, innovative thinking among other competences. In order to facilitate students with the competences expected in professions, educational institutions worldwide are implementing innovations and changes in many aspects.

This book series include a list of research projects that document innovation and change in education. The topics range from organizational change, curriculum design and innovation, pedagogy development, to the role of teaching staff in the change process, students' performance in the aspects of not only academic scores, but also learning processes and skills development such as problem solving creativity, communication, and quality issues, among others. An inter- or cross-cultural perspective is studied in this book series that includes three layers. First, research contexts in these books include different countries/regions with various educational traditions, systems and societal backgrounds in a global context. Second, the impact of professional and institutional cultures such as language, engineering, medicine and health, and teachers' education are also taken into consideration in these research projects. Thirdly, individual beliefs, perceptions, identity development and skills development in the learning processes, and inter-personal interaction and communication within the cultural contexts in the first two layers.

We strongly encourage you as an expert within this field to contribute with your research and make an international awareness of this scientific subject.

For a list of other books in this series, www.riverpublishers.com
http://www.riverpublishers.com/series.php?msg=Innovation_and_Change_in_Education_-_Cross-cultural_Perspective

School Culture Improvement

Zhang Dongjiao

Published, sold and distributed by:
River Publishers
Niels Jernes Vej 10
9220 Aalborg Ø
Denmark

River Publishers
Lange Geer 44
2611 PW Delft
The Netherlands

Tel.: +45369953197
www.riverpublishers.com

ISBN: 978-87-93237-92-6 (Hardback)
978-87-93237-93-3 (Ebook)

©2015 River Publishers

All rights reserved. No part of this publication may be reproduced, stored in a retrieval system, or transmitted in any form or by any means, mechanical, photocopying, recording or otherwise, without prior written permission of the publishers.

Contents

Foreword

Schools are cultural education institutions which are designated to spread knowledge and cultivate talents. Therefore, schools must attach importance to cultural construction and create an educational environment to enable students to lively and initiatively develop at school. Excellent school culture isn't naturally formed; instead, it requires all students and teachers to create it heartedly for there are distinctions between advanced and backward culture as well as high and low culture. If a school emphasizes on the enrolment rate instead of students' character education and personality development, the culture of this school is utilitarian and backward; if teachers of a school lack enthusiasm in caring and loving students, the culture of this school is low and even vulgar. To construct the advanced school culture needs the efforts of all students and teachers.

School culture is the soul of a school, which contains the common values, belief, wish and efforts of all teachers and students. Therefore, school culture plays a leading, standardizing and incentive role.

School culture consists of spiritual culture, systematic culture, school material culture, students and teachers' behavioral culture. As its core, spiritual culture reflects the school-running philosophy, educational concept, values and way of thinking. School-running philosophy and educational concept are first shown in the outlook on talents, namely how to cultivate talents. Schools are supposed to be education-oriented to cultivate the socialism builders and successors who develop morally, intellectually, physically and aesthetically, pursuant to the national educational policies. Schools' spiritual culture is also reflected in the outlook on students as well as the relationship between teachers and students. Teachers should develop such a concept that all students will become state pillars and should love every student without any bias; the relationship between teachers and students shall be equal, democratic and harmonious. With such a teacher-student relationship, education can move on successfully. School culture is also reflected in the curriculum and instruction. Curriculum and instruction are the carriers of culture. During the instruction, teachers are required to stress on the cultural connotation, values, thoughts and emotions except imparting knowledge.

Schools' systematical culture is also significant. School affairs are relatively complicated and heavy, including instructional work, ideological work and logistical support. Schools are places where crowds congregate. Inevitably, there will be various conflicts and contradictions. Schools shall have a certain systems to tell students and teachers dos and don'ts and individual responsibilities. The establishment of systems shall be associated with and guided by the school-running philosophy. That is to say, the establishment of systems shall conform to the spiritual culture construction of schools.

Schools' material culture construction includes school building construction, school design and landscape layout, etc. It not only guarantees the educational and instructional work but also reflects schools' spiritual outlook. The construction of schools' material culture shall be especially students-oriented and shall reflect the dominant culture of schools and make students and teachers feel comfortable and happy and thus willing to study and live in such an environment. Schools shall highlight the ceremonial and symbolic construction. A piece of school motto, a school badge and a school song often manifest a school's spirit. In a word, every part of the school shall be of educational significance.

It's impossible to develop school culture in a short time. School culture construction needs the active participation, joint planning and efforts of all students and teachers. School culture improvement is a process to construct the school culture from the perspective of management, which can bring vigor and vitality to school culture.

Professor Zhang Dongjiao has been engaged in the educational management research for a long time. She once was the Dean of College of Education Administration of Beijing Normal University. Under her leadership, a batch of young and middle-aged scholars went to grassroots schools and compiled this book. From the perspective of management, this book expounds the construction of school culture in the aspects of theories, practice, concept and operation. The publishing of this book must be beneficial to grassroots schools to improve their school culture and meet the requirements of "well running every school and well educating every student" proposed in *National Long-Term Plan for Educational Reform and Development (2010–2020).*

Written in Beijing Qiushi Bookstore on August 19
Prof. Ming yuan Gu, Beijing Normal University

Acknowledgement

After years of experiments, research and writing, and under the elaboration of RIVER PUBLISHERS *School Culture Improvement Series* finally comes out. There are many people directly or indirectly involved who are worth being appreciated and remembered.

My thoughts were left in the tens of schools I walked through, and so were the thoughts of my colleagues who are like my brothers and sisters. This series won't come out without them. I don't know why, when speaking of them, I always recall Zhao Benshan's lyrics: many things have happened in our village, and I think they are so funny when I recall them. I'd like to introduce those people in our village. It may due to my birth in a village that I like these earthy words and real-life reasons. They are young and strong, with their common ground of brightness in their heart, good literary talent and high intelligence quotient; and they are enthusiastic, kind, broad-minded, responsible, conscientious and gregarious. Now they are respectively busy, so I'll silently think of them, from the old to young.

Zhou Wei is a researcher. He and I are cross-age friends. With penetrative voice and preciseness, he has numerous ideas and extensive knowledge. He is the managing director of the project. Mr. Zhou makes me feel relaxed and have someone to rely on. When he is absent, I'm the "head" and I have to be concentrated. Professor Ma Jiansheng is a Comparative Education Doctor. He loves on-site challenges and extemporaneous lectures when teaching. He is honest and often gives great ideas with deep insight. Doctor Gao Yimin is engaged in comparative education. He is gentle, humorous and breathtakingly handsome. He cares about international events and knows everything from Obama to Gaddafi. Yu Kai has gotten two PhDs in China and U.S. respectively in comparative education and educational leadership. He is careful and dainty, with a little bit melancholy and romance. Xu Zhiyong is a Doctor of Educational Management. He is practical and receptive, amiable and humorous. He is good at quantitative research. Zhao Shuxian is a Doctor of Educational Management, while he majored in physics in university. He is good at quantitative research and also loves poems, just like a literary young woman. Doctor Yu Qingchen is the youngest, born after 1980s. He is a prince of philosophy with wisdom and idea filled in his head. He is sedate and intelligent and he is also a good talker. Next to Mr. Zhou, I'm the oldest. After the broadcast of TV series *My Chief and My Regiment*, I have been promoted from team leader to "chief". I'm not arrogant for this. I think I'm just like a hen, finding food, gearing up and protecting them. In addition to these partners who always work together, Doctor Wang Chen and Doctor Hong Chengwen from Beijing Normal University as well as Professor Chen Li from Beijing Institute of Education have participated in this project for one or two years, but they left due to going abroad or being busy. There are also many PhD candidates and Master's candidates who once participated in, totaling nearly a hundred. They are all intelligent and whoever hired them are lucky. We've also cooperated with seventy or eighty principals, who are all renowned, shrewd and capable, and you can see their names in the book or contents. Besides, Commission of Education is really nice. They implement policies and ensure funds, and are the "matchmaker" between schools and us. They supplied the best service. The standardization construction of primary schools in Fengtai of Beijing was divided into three stages which cost three years. In these three years, the project was conducted under the leadership of Director Feng Xiaoguang, three deputy directors including Li Yongsheng, Zhang Lixin and Liu Jianbin, three chiefs of primary school education section and Dean Zhu Shicheng of Fengtai Branch of Beijing Institute of Education. Deputy Director Zhang Fenghua and Section Chief Wu Jin would come to the site personally every working day of Haidian Project. In the initial stage, Zhang Yingyu was the contact person of Haidian Project with Commission of Education. We all like her as she is slim, tall and pretty, and more importantly, she is a good talker and she is prepared to take the initiative. In addition, those friends of the magazine *Management of Primary and Secondary Schools*, Chief Editor Sha Peining, Director Sun Jinxin and Editor Xu Liyan, laboriously followed us in conducting activities, and it's really rare that we always enjoyed it. Those people mentioned above are leading roles, live and vivacious and full of energy. They deserve to be appreciated and eulogized. I hereby give them verbal praise.

There are two more special "persons" who need to be solemnly appreciated. They are my teachers, who are elder than me and are "big bugs" who are influential in education. One is Professor Gu Mingyuan, who claims to be "after 1980s". He lives a secular life but he is not bound by conventions, and he is earnest in supporting juniors. When he was 80, he attended the kick-off meeting for Fengtai Project, which was held by me. At 82 years old, he was invited to the summing-up meeting. Together with Mr. Gu, another teacher of the similar age with us but with much greater achievements, Professor Zhang Binxian, was also invited to the summing-up meeting. He was my teacher of foreign education history when I was an undergraduate. He is amiable and very knowledgeable and his lessons are attractive. Both of them are advisors of the Research Institute of Improvement and Development (renamed as School Culture Research Center after the establishment of Faculty of Education) in our school, so I often shamelessly invite them and consult them. I'd like to express my gratitude to them and I also know they will continue supporting me.

Never forget the well-diggers when drinking from the well. Thank the Ministry of Education for approving this project and funding it. This book series is the main achievement of Humanities and Social Science Project of the Ministry of Education "Research on the Model of Regional School Overall Development Driven by School Culture" (2010–2013), New Century Outstanding Talents Project of the Ministry of Education and 985 project of Beijing Normal University "Building and Application Research of the Model of School Overall Development in the Region Driven by School Culture" (2011–2014).

Thanks for the wisdom and labor of those graduate student. Thank RIVER PUBLISHERS for taking care of *School Culture Improvement Series* and providing such a good display platform and we pay tribute to XIANGYUN DU who contributed his intellectual effort. Your meticulous and professional dedication and superior and sagacious revision suggestions are amazing. You are the protector and creator of the quality of education publishing. Thank the authors of those references giving nutritional support to this series. You're predecessors of this research. We pay tribute to you. We're very glad and expected to hear the ideas and opinions of education colleagues on any book or paragraph. The world is vast but the field is not, and we will meet one day.

School Culture Improvement leads schools to break the glass roof of experience management and open the door to the rational world. With a high intention, schools have better goals, few limitations and more strength and they won't fall. My colleagues and I frequently visit schools to "learn from them", and after we come back, we study and digest those in the ivory tower and continue progressing ceaselessly. Nobody dare say their cognition of school and school culture is the same as that a few years ago. It's lonely at the top, but people won't fall from the "top" so soon. That is to say, our colleagues in universities, secondary and primary schools are "flying" steadily with culture. If so, invisible wings are real.

"Culture is a long way to go but culture is also within our sight", this is the slogan of *Cultural Perspective* program. I'm obsessed with this sentence, for it's elegant, charming, profound and philosophic. Management of culture is such a laborious work. Management of culture sounds hip and superior but actually is extremely ordinary. Plainly speaking, it means putting human in the center of education, school, management and in minds of managers. Education wins due to philosophy and schools win due to culture. My companies and I are walking on the road of believing, practicing and verifying it, so the happiness and growth we've earned is in the past, at present and also in the future.

Earnest makes my words fluent and smooth; and I still have a lot to say but the daybreak is coming. This is a postscript for memorizing a research and a kind of growth.

1

Overview of School Culture and Its Management

School culture is the power to promote sustained and stable development of school and, it is the only way for school to get cohesion and competitiveness and to build learning community. School culture improvement makes school step from experience management and scientific management into a higher cultural management stage, so as to ensure that school education realizes the man-centered education ideal. This chapter interprets and presents culture value orientation for running schools and its abundant practice in the perspective of school culture connotation, school culture structure and school culture improvement

1.1 Connotation of School Culture

1.1.1 Analysis of School Culture

Culture is a way by which human impacts the power of surroundings. One of the most important ways is school education. School is a social organization. Terrence E. Deal and Allan A. Kennedy who are famous for their research on enterprise culture believed that organizational culture is man's way of doing things in an organization. The culture consists of enterprise environment, values, heroes, rituals and ceremonies and cultural network, among which values are the core of organizational culture.[1] William Ouchi pointed out that corporate culture consisted of its tradition and ethos. The Culture includes the values of a company, such as aggressiveness, defensive, flexibility, namely the values determining activities, opinions and actions.[2] The definitions of organizational culture are all focused on the compositions and the core of the organizational culture. The culture is not only concrete but also valuable. The core of organizational culture is the values shared by organization members and the philosophy to guide people's behavior. In order to define school culture, we need to start with development of its concept and a few relevant concepts.

1.1.1.1 Campus culture

In the development process of school culture concept and its practice, campus culture occupies an important position. In the middle of 1980s, China mainland started using the concept of "campus culture" in students' activities in the universities. There are two landmark cases in practice: in November 1986, Shanghai Jiao Tong University held the "Symposium on University Campus Culture in Shanghai"; in April 1990, China Association of Higher Education, The Chinese Society of Education and the Propaganda Department of the Central Committee of Communist Youth League jointly held "Theoretical Symposium on National Campus Culture" in Beijing.[3] The research of the campus culture is distinctly divided in two stages: in the initial stage, in both practical and research areas, the campus culture was almost the same with the campus spirit, or it was equal to the art education or cultural activities; thus the research of the campus culture was limited to the art education and club activities. With the enrichment of practice and the deepening of thinking, the research on campus culture stepped into the second stage, namely the campus culture became the synonyms of the school culture. The club activities and campus atmosphere still accounted for a significant proportion, but they were not the only aspects the research was focusing on. Scholars and schools

---

[1]Deal and Kennedy. Corporate Cultures: the Rites and Rituals of Corporate Life [M]. Translated by Tang Tiejun, Ye Yongqing, Chen Xu. Shanghai: Shanghai Science and Technology Press, 1989: 12.

[2]Ouchi. Theory Z: How American Business Can Meet the Japanese Challenge [M]. Translated and Revised by Sun Yaojun, Wang Zurong. Beijing: China Social Science Press, 1984: 169.

[3]Yu Qingchen. School Culturology [M] Beijing: Beijing Normal University Press, 2010: 1.

encourage and support to alter the research from campus culture towards school culture, since there are distinctions between the two.

"Campus" refers to a spatial dimension area, or a place, while "school" refers to an organization to nurture and educate people. Campus is one of the parts of school; it's more about showing a cultural concept of static, school-entity type. Zhao Zhong, one of the scholars focusing on the school culture research, pointed out that the school culture was a community culture, of which the main space is the campus, the main subject is the student, and the main character is the campus spirit. The school culture possesses integrity; it's the school features, regulations and spiritual atmosphere which the school subject has developed with unique cohesion. The core of school culture is common values formed in the long-term process of running school. Campus culture usually refers to the students' activities, while school culture covers the aspects of teacher culture, student culture, course culture, organizational culture and environment culture. Campus culture is a subsystem of school culture, and it's more convenient for further discussion about the issue under the topic of school culture.[4] Then how can we embody the dynamic practical process of the school cultural construction? China has started to attach importance of the school culture's effect in school development, consistent with the trend of the world education development. *The Education Culturology* by Zheng Jinzhou, for example, discusses school culture and subculture from the view of organizational studies and culturology; Fan Guorui, in his book *The Theory and Practice of School Management*, discusses school organizational culture with a separate chapter. Meanwhile, the education practices in K–12 schools still use the concept "campus culture" a lot, but the connotation and practice have already followed the direction of school culture. Indeed, the school culture reflects the internal quality and developing trend of a school better than the campus culture does.

1.1.1.2 School spirit

School spirit is closely related to school culture. School spirit, in the long-term education practice, is the dominant consciousness of school, which is formed in careful cultivation for development, and combined with school individuality. Every school has its own distinctive culture spirit, which is usually generalized through concise and rich philosophical language. It can also be expressed by the forms of school song, school motto, school badge, etc. School spirit should

---

[4]Zhao Zhongjian. School Culute [M]. Shanghai: East China Normal University Press, 2004: 95.

be practical and realistic.[5] School spirit means cultural spirit of school, which is the cohesion of school culture and concise expression of school culture characteristics.

1.1.1.3 School atmosphere

There are differences between school atmosphere and school culture. School culture explores school from the view of sociology and anthropology and in the process of research the method of qualitative research is mostly used. It tends to use the method of deep description to understand how school transmits its symbolic significance. School atmosphere explores school from the view of psychology and in the process of research the method of quantitative research is mostly used. It tends to use the method of investigating and multivariable statistical analysis to ensure the types of behavior affecting the development of schools. Compared with abstraction of school culture, the research of school atmosphere is more specific and its operability is stronger.[6] School culture and school atmosphere are two overlapping concepts. School atmosphere, to a large extent, affects school culture improvement. The types of school atmosphere and those of school culture are highly correlated.

Kim S. Cameron and Robert E. Quinn's distinction between organizational culture and organizational climate has greatly inspired us. According to their research claims, it can be inferred that school culture is different from school climate. School culture is a series of assumptions of enduring values, belief, and expression for schools and the characteristics of their members; school climate refers to brief attitude, feeling and personal understanding. While school culture is a lasting core quality changing slowly, school climate is based on attitude, which changes swiftly and dramatically. School culture is internal organizational characteristics that are difficult to identify while school climate is more obvious and visible organization characteristics. School culture contains core value and consensus to things while school climate contains situational changes and changeable personal views when meeting information.[7]

---

[5]Zhao Zhongjian. School Culture [M]. Shanghai: East China Normal University Press, 2004: 307.

[6]Miskel & C. G. Educational Administration: Theory, Research and practice [M]. New York: McGraw-Hill, Inc. 1996: 161.

[7]Kim S. Cameron, Robert E. Quinn. Diagnosis and Change of Organizational Culture [M]. Xie Xiaolong, Translated. Beijing: Renmin University of China Press, 2006: 110.

1.1.1.4 School culture

School culture is a spiritual guiding force, a characteristic system and behavior. The first one to raise the concept of school culture was the American scholar Willard Waller, who used the expression "school culture" in *Sociology of Education* in 1932: particular culture formed in schools. This kind of culture is that children of different ages change adult culture into simple forms or keep adult culture through children play community; and teachers design and guide students' activities to stimulate the culture to come into being.[8] Paul E. Heckman deemed school culture to be a belief which the headmasters, teachers and students hold and their behaviors are dominated by; school culture and school tradition are closely related to the history.[9]

Educational Dictionary, edited by Gu Mingyuan, defines school culture as: values and behaviors related to instruction as well as all other forms of activities at school.[10]

Zhao Zhongjian pointed out that school culture was the combination of standard and traditional values, thoughts and behaviors. Its core was the management philosophy and the school spirit, and its exterior manifestation was the school manner.[11] Zheng Jinzhou held the view that school culture was the concepts and behaviors all members or some members commonly acquired and possessed.[12] On the basis of summarizing numerous research findings, Yang Quanyin and Sun Jialin thought school culture contained two parts: the interior part was the values; the exterior part was presentation forms including behavior standards, ceremonies, visual symbols, etc. Among them, core was the values, and presentation form was outer case; values were the fundament, and presentation form was of little significance.[13]

Wang Jihua came up with a concept of educational new culture. He deemed that educational new culture is a dynamic organic whole composed of nine cultural forms, namely educational value orientation as the core, headmaster

---

[8]Zhong Qiquan. Essentials for New Curriculum Faculty Training [M]. Beijing: Peking University Press, 2002: 100.

[9]Heckman. School Restructuring in Practice: Reckoning with the Culture of School [J]. International Journal of Educational Reform, 1993 (2): 263–272.

[10]Gu Mingyuan. Educational Dictionary: Volume 6 [M]. Shanghai: Shanghai Education Press, 1992: 426.

[11]Zhao Zhongjian. School Culture [M]. Shanghai: East China Normal University Press, 2004: 99.

[12]Zheng Jinzhou. Educational Culturology [M]. Beijing: People's Education Press, 2000: 240.

[13]Yang Quanyin, Sun Jialin. School Culture Research: Cultural Perspective on a Middle School [M]. Beijing: Educational Science Publishing House, 2005: 46.

culture as the commander, teacher culture and student culture as the man-centered philosophy carrier, static campus culture as the physical carrier, kindergarten culture, class culture, dynamic campus culture and school culture as the inheriting carrier and examination culture as the innovative impetus. School culture is a cultural field demonstrating various educational culture forms and reflects the direction of school development. School culture macroscopically guides the development direction of headmaster culture, school educational orientation culture, teacher culture, student culture, dynamic and static campus culture, class culture and examination culture. The operation of school culture should be materialized through particular carriers, the design of which embodies the cultural strategy of school development. School culture contains six carriers: environment carrier, concept carrier, activity carrier, instruction carrier, system carrier and behavior carrier.

Given all that, the research suggests that the school culture is an organizational culture and the combination of the behaviors and the manners, as well as the physical forms of all members dominated by schools' core values, including spiritual culture, systematic culture, behavioral culture and material culture. Among them, values are the core and the soul of the school culture. School culture can be established and grow, and be condensed and illustrated as well. But the school culture defined academically cannot be met in reality, and cannot be adopted mechanically. What the researchers and practitioners should keep in mind is that the school culture is alive, motile, contextual and factual, instead of exoteric and conceptual. Managers need to construct, nourish and cultivate school culture according to context and reality, and researchers need to explore and learn school culture in context and reality. Initiative person can form school culture through changing and reforming environment, and can promote environmental change and school development through managing school culture.

1.1.2 School Culture Type

School culture is divided into different types according to different standards. Considering the effect factors to form school culture, school culture is closely related to such factors as school administration mode, organizational atmospheres, and leadership style. Different schools have different administration modes. Tony Bush proposed five administration modes and their characteristics: conventional administration is to realize school goal by hierarchical administration in a rational manner; democratic administration is to advocate teachers to share powers and policies; political administration is to bargain before decision making; subjective administration is to encourage to play a role of individual to realize their goals; and fuzzy administration's decisions are made by different committees and working group. Different administration modes form different organizational atmospheres, so we can grasp the school culture types and their characteristics through paying attention to the types of school atmospheres. A.W. Halpin and D.B. Croft put forward four organizational atmospheres: open-end organizational atmospheres with close relationship, mutual cooperation and respect among teachers, close-end organizational atmospheres in lack of commitment and with mutual disinterest, busy organizational atmospheres under which headmaster regulates the behaviors of teachers to enable them to perform at a high level, and loose organizational atmospheres under which headmaster provide supports to teachers actively, but teachers cooperate less with each other and lack commitment. Therefore, different organizational atmospheres form different school cultures. The school culture classifications put forward by Robert E. Quinn and Michael McGrath are as shown in Table 1.1.

School culture is divided into growth, mature and senescence school culture in accordance with the establishment process and development stage. It is also divided into such sub-cultures as headmaster culture, teacher culture, student culture, official subset culture, unofficial subset culture in accordance with different subjects. The study on the types of organizational culture provides us with a useful framework to analyze the school culture type.

Table 1.1 School culture classification and their characteristics

School Culture Type	Value Orientation	Basic Characteristics
Rational culture	Efficiency and achievement	Centralized, conventional administration mode + busy organizational atmosphere
Hierarchical culture	Clear responsibilities	Centralized, conventional administration mode + close-end organizational atmosphere
Contractual culture	Participation in discussion	Decentralized, democratic administration mode + open-end organizational atmosphere
Developmental culture	Innovation changes	Decentralized, fuzzy administration mode + open-end organizational atmosphere

"Open-control type" school atmosphere put forward by A.W. Halpin and D.B. Croft connects behaviors and manners of headmaster and teachers, from open to close, including open, autonomous, controlled, free, paternalistic and close-end school atmospheres. The analytical framework of "healthy-morbid" school atmosphere of school organization put forward by Wavne K. Hoy et al focusing on the health of interpersonal relationship between students, teachers, administrators and community members divides school into healthy school and morbid school. Willard Waller holds that school culture includes two basically opposite cultures, including teacher-directed adult society culture and student-directed group culture. There are many cultural conflicts between such two school cultures, which always end up with the victory of the adult culture and the socialization of students. It is an effective way to inspect school culture through the mode of student control. Conway Edel designed student control continuum from regulation to humanism, putting forward regulatory and humanistic school atmospheres. Likert put forward four regulatory and matching organizational atmospheres in accordance with the characteristics of leader-member relation in organization: severe and authoritative type, merciful and authoritative type, consultation type and democratic participation type.

Carl R. Steinhoff & Robert G. Owens consider that the organizational culture shall be defined from six aspects, namely organization history, values and beliefs, myths or stories on organization, cultural norm of organization, and traditions, rituals and ceremonies showing organization characteristics, and the heroes of organization. The put forward four kinds of public school cultures:[14]

Family culture: This kind of public school may be represented by family or group, for example, headmaster and teachers are described as parents, nursery governesses, friends, partners or couches. In such a school, it is most important to care about each other, and everyone is glad to be one of the big families and contribute to it. They care for all students, full of friendliness and cooperation.

Machine culture: This kind of public schools may be represented by machines, such as the machine full of fuels, political machine, busy apiary, and rusty machine. School is compared to machine in terms of the instrumental significance. Machine's driving force is derived from the tight structure of organization. The description of administrators is based on their capacity to maintain input, for example, headmasters are described as workaholics,

[14]Quoted from Lunenburg & Ornstein, Educational Administration: Theory and Practice [M], Trans. Liu Zhijun et al, Beijing: China Light Industry Press, 2003: 63–64.

lumberjacks and generals. The public school has a tight social structure. Different from family school culture, the aim of such schools is to provide protection but not loves for members and school is the machine for teachers to finish work.

Performance culture: This kind of public school may be represented by circus, the Broadway Melody, banquet and elegant ballet performance. The headmaster is regarded as the master of ceremony, tightrope walker, and the leader of circus. Similar to the teachers in family culture school, teachers of this kind of public schools also experience the mutually dependent social activities, with a difference that it is the relation between performance and feedback in such culture. This kind of public schools has a high requirement on artistic quality and wisdom in teaching.

Culture of horrible scenes: This kind of public school may be represented by nightmare about wars and revolution, unpredictable and full of tension. Here it is difficult to know the next underdog. Teachers name the school as a close box or prison. The headmaster is like a statue with an automatic cleaning function, ready to keep his position anytime. In this kind of schools, the main function of administrators is to cope with everything, with full domination and control, and employees should adjust to them. Different from family culture and performance culture, teachers in this kind of public school live a lonely life, participating in fewer social activities. The organization hopes everyone to abide by rules and regulations, and smile at a proper time. Employees always use foul language, so there is no close relationship among them. It is an indifferent, hostile and skeptical culture. Everything is possible, as is often the case.

According to the completeness, clearness and strength of school culture system and learning from research method and achievement of spectrum, this study put forward the school culture spectrum theory.

The spectrum is a pattern formed according to a light wavelength (or frequency) in sequence when polychromatic light passes through a dispersion system (raster or glass prism). Based on its formation nature, spectrum may be divided into emission spectrum, absorption spectrum and dispersion spectrum. The emission spectrum may be divided into line spectrum, band spectrum and continuous spectrum. The line spectrum, also known as atomic spectrum, is mainly generated from atom, consisting of some discontinuous light line. The band spectrum, also known as molecule spectrum, is mainly generated from molecule, consisting of some tense light with a certain wavelength range. The continuous spectrum is mainly generated from the emission of electromagnetic radiation of stimulated incandescent solids, liquids, or high pressure gas. The continuously distributed spectrum from red light to purple light consists of continuously distributed spectrum in all wavelengths. Since each atom has its own characteristic spectral line, we can identify the material according to spectrum and determine its chemical composition, which is known as spectrum analysis. The spectrum of different elements has different positions and spectrum line in different colors, or lacks some spectrum lines. However, the spectrum line of material containing the same elements is always at the same position. Spectrum analysis is to analyze the element composition of material through this principle.

Culture spectrum analysis is a unique and proper study and expression vision. School is a culture existence, so every school is a unique culture scene—nothing good or bad about the school culture. Every school has its own culture, and there must be differences in strong and weak culture. The school culture spectrum theory is the mature model of school culture, mainly involving several aspects below.

Firstly, school culture zone. The school culture spectrum theory divides the school culture type and modality into linear culture zone, strip culture zone, and continuous culture zone from weak to strong in accordance with its maturity. Those are three culture zones, and the transmission mark between different types is the school culture spectrum which is the completeness and clearness or separation status of school culture, or uniformity of data point, etc. Figure 1.1 is the school culture zone spectrogram.

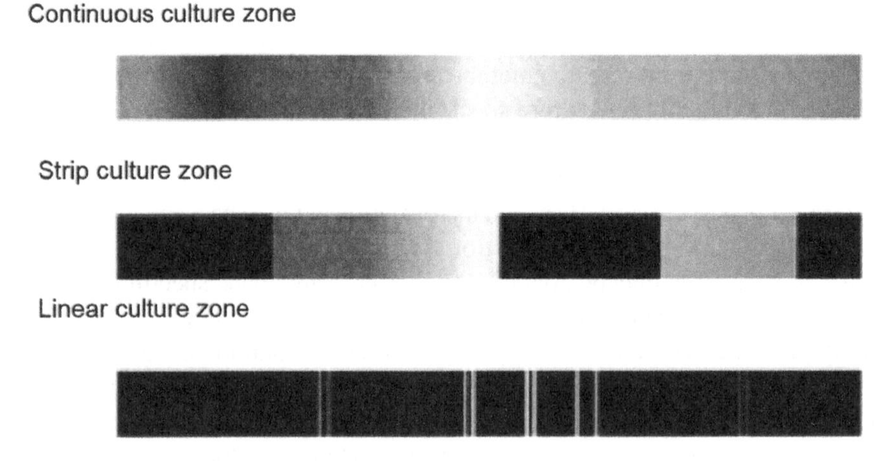

Figure 1.1 School culture zone spectrogram.

In our studies, school clusters sharing the same data point and located at continuous culture zone are known as school groups with strong cultures while the schools with data point at strip culture zone are known as school groups with medium cultures, and the schools at linear culture zone are known as school groups with weak cultures. There are several schools in each culture zone, with different culture strength statuses which consist of many data points.

Taking the school with strong culture for example, the characteristics of its culture spectrum include: a complete school culture system, including core value system and practice system; consistent logic—with an integration of theoretical and expression logic of form and content, with multivariate existence; uniform data point. For instance, the spectrums with equal wavelength form spectrum line, determining the position of the school in culture spectrogram. The data points of the school with strong culture shown in the spectrogram are that it has a clear core value which is recognized and supported by members; it has a diversified, harmonious and uniform educational philosophy system which is stated clearly, uniformly, scientifically and reasonably; it has a complete educational practice system which highly supports and bears core values from systems, behaviors to activities and physical forms; it has positive heroes and models who are recognized by everyone; it seeks to work perfectly and wear properly, carrying out etiquette training regularly; it is filled with friendliness and unity; the school's meeting focuses on teacher and students' development, seeking for efficiency; and it carries out activities with a typical ceremony, characteristics and features. However, the characteristics of the school with weak culture include that it lacks specific value; or has many values or thoughts which are stated confusedly and lack consensus; or it has different belief systems in different departments; or it has no heroes or sets disintegrative examples; or has different dressing and conversation standards; or it holds meeting draggingly and ineffectively; its employees always expose discontent; it has many unofficial subsets; and the rituals of everyday life has no organization or are contradictory. Such signs are the indications of school culture dangers.

Secondly, cross section spectrogram of school culture. "Birds of the same feather flock together", so there are many classification problems in natural science or social science. For school clusters at the same time in different spatial location, there is clustering phenomena in school culture types. The cluster refers to the process to divide the set of physical object or abstract object into several classes consisting of similar objects. The clusters from clustering are a set of a group of data objects which are similar to each other in the same clusters and different to the objects in other clusters. School culture spectrogram consists of linear culture zone, strip culture zone and continuous culture zone. Its characteristics are to arrange school culture from the weak to the strong or from the strong to the weak by means of clustering analysis in order to facilitate the research. So to speak, it is a cross section static spectrogram of school culture. There are different school culture spectrograms in different times.

Thirdly, vertical section spectrogram of school culture. For cultural spectrogram, its smallest unit is school. Each school will find its own spectrum line and position in the cultural spectrogram. For the culture of a school, how does the school culture spectrum theory describe the composition of the smallest unit? At different stages, among different members and in different spaces, the culture development status of each school is inherited and continuous, but different. That is to say, from the perspective of the development history, the culture development status of every specific school will fluctuate, which is not always a process of linear development from weak to strong but nonlinear development. It is possible that it fluctuates from strong to medium and then to weak, or from weak to medium and then to strong, or from medium to weak or strong; or remains the same. It is vital of the influence of ideology and mainstream value, reformation of school organization, different leadership styles, changes of school address at various times. It seems that there are changes of culture type and spectrum line in a school at different times and spaces, and the changes don't develop uniformly, but the uniform velocity can be increased by control. From the perspective of historical development, the vertical section spectrogram of school culture is a dynamic historical drawing. (See Figure 1.2)

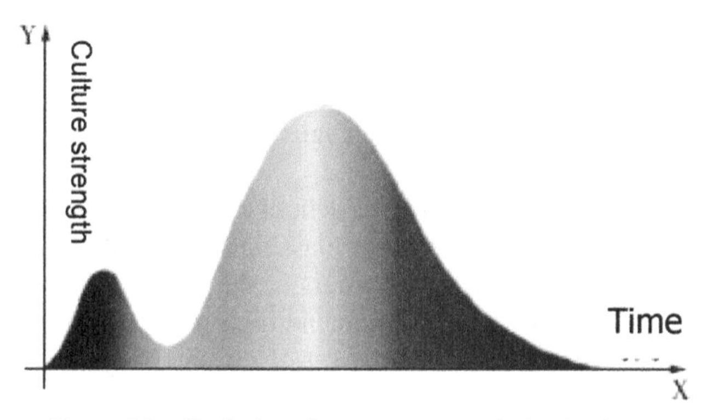

Figure 1.2 Vertical section spectrogram of school culture.

Fourthly, diversified school culture goals. In the school clusters at the same time and space, there are various culture development status, and the culture improvement and construction goal of each school are full of variety. Each school may develop improvement goals according to the starting point of its culture data point. The school whose spectrum line and position are in linear zone has to take efforts to move toward medium zone; the school whose spectrum line and position are in medium zone has to take efforts to move toward continuous zone, and the school whose spectrum line and position are in continuous zone may also take efforts to move toward strongest zone.

In the process of school culture improvement, it is unnecessary to stick to the specific description above but know about these types based on the actual development of school, and provide a good reference to the value orientation and development direction of school culture improvement.

1.1.3 School Culture Development Stages

With reference to the revolution stages of organizational culture and the researches on school culture development stages by researchers, school culture development can be divided into four stages within a cycle.

1.1.3.1 Initial stage

The initial stage refers to the initial stage of schools or the initial stage of school culture revolution when schools are in the linear culture zone and this stage needs 1 to 2 years. There are three characteristics of school culture development in the initial stage:

First, the headmaster is cautious and dominant. There is a view that school culture, i.e. headmaster culture, usually reflects the headmaster's value and leadership style and the headmaster's knowledge and taste usually lay a foundation for the initial substance of school culture. The headmaster plays a core leadership role for the school culture establishment and improvement. If a headmaster newly comes to a school, or he conducts culture revolution in the original school, he needs to be cautious in playing this dominant role: He should respect the history and tradition of the school, and start to work based on sufficient investigations and scientific diagnoses.

Second, the headmaster needs to gather suggestions and opinions from all levels of schools based on investigations and diagnoses, lead school members to explicitly put forward core values of schools after preliminary summaries and repeated revisions, try to infuse the organization with it and actively spread it to teachers and students. In schools with strong cultures, the headmaster needs to think how to conduct the revolution on the basis of knowing about a large amount of information thus to make school cultural concept system and practice system more coordinated, perfect and excellent.

The third is centering on core values of schools to preliminarily conduct systemization and formulation of regulatory framework. During this process, imitation may be dominant. In this stage, school members are still not familiar with the new headmaster or the new thoughts of the headmaster to conduct the culture revolution, so most of them are still taking a wait-and-see attitude. Therefore, school culture improvement is quite like "one individual's battle" of the headmaster and the headmaster is a "lonely thinker".

1.1.3.2 Growth stage

After schools experience the initial stage of the establishment, they quickly enter into a growth and development stage by leaps and bounds. Schools are transitioning from the linear culture zone to the strip culture zone and this stage needs 3 to 5 years. There are three characteristics of school culture development in the growth stage:

First, the headmaster serves as the middle leadership. The headmaster has already been familiar with all the work and all the members of schools and preliminarily laid the foundation for the leadership authority, and can stand in the middle to command, help and guide followers.

The second is that school culture construction team forms preliminarily. The leadership team centered with the headmaster has been stabilized preliminarily with the same objectives. During this stage, schools will consider and start formulating a development plan and clear expectations in the future 3 to 5 years and confirming school-running objectives and cultivation objectives of schools. With the formulation of the development plan as tongs, the decision-making cores will be reduced. Moreover, all the faculty members take part in it and propose suggestions and opinions regarding the school development plan and departments' plans, as well as continuously publicize, revise and generalize the core values.

The third is to perfect the regulatory framework of schools on the basis of widely soliciting opinions to make it more reasonable and serve the development of people and schools. During this stage, the encouragement for people and the establishment of system are the emphases and a variety of standards start to operate stably. Under the passionate leadership of the headmaster, all the members of schools are in their proper place and all are involved in the battle. The headmaster has become a "standard-bearer" and "coach" full of passion.

1.1.3.3 Stage of ripeness

After the regulatory stage, school culture development enters into the stage of ripeness. School culture is in strap structure, and this stage needs 2 to 4 years. There are four characteristics of school culture development during this stage:

First is the calm advancement by headmasters. All the members in schools with strong culture are motivated fully and authorization becomes a pleasure for the headmaster. The subjective influence of leaders turns less dominant and direction control, resources and conditions provision and systematical rethink about the school culture have become the major tasks. The culture stereotypes precipitated for a long time will lead the school development. Such conversion for the headmasters in schools with weak cultures will be achieved a little later, and those headmasters may even always stand on the stage.

Second is the implementation of the development plan centered on school core values, and the emphasis on the interpersonal culture and behavioral culture construction. The development plan and departments' plans are disintegrated and allocated to each position, and responsibilities are firmly linked with each faculty member. The headmaster leadership team's behaviors, middle management team's behaviors, teachers' behaviors and students' behaviors have their own norms. What's more, they should have personality of the schools. Meanwhile, it is necessary to further refine on the design of activities and ceremonies.

Third, there have been heroes in schools and the examples of teachers, students, leaders and support staff have been set up and they explain the school values internally as well as represent school culture standards and achievements externally. The combination of the advanced things with examples can greatly promote the school culture construction. If values are the soul of culture, heroes are the humanized reflects of these values and the concentrated shadows of organizational power. In schools with strong cultures, heroes are in a key position, and they are the compasses. It tells us that heroes are beside us and the success is in our power.[15]

[15]Terrence E. Deal and Allan A. Kennedy. Corporate Culture: The Rites and Rituals of Corporate Life [M]. Translated by Li Yuan and Sun Jianmin. Beijing: China Renmin University Press, 2008: 37.

Fourth, schools start material culture construction. Based on the clear core values and the school-running practice thought, the design of visual-audile identification system is started, expressing and enriching the school spirit and culture. In the stage of ripeness, school culture is fully hierarchical and all kinds of systems from spiritual culture and system culture to behavioral and material culture have all been formed. After formation of school culture, school members water and plant it together and absorb new components to develop an open-end and developmental culture system.

1.1.3.4 Compact stage

If the first three stages of school culture development are future-directed stages, then the compact stage is the review stage in the process of progress. After stable characteristics are formed in school culture, it is necessary to further compact and promote it. During the compact stage, continuous characteristics are reflected in school culture, data points are gradually supplemented and perfected, and there are three characteristics as follows reflected in school culture development:

The first is the headmaster's contingency leadership. The headmaster of schools with mature culture will choose and adopt the contingency leadership with ease according to the changing circumstance. "Contingency" means to be adaptable to changing circumstances. The contingency leadership contains three points. In the sense of time, it refers to changes of ways and means in management caused by the passage of time and changes of school environments. In the sense of space, it refers to changes of ways and means in management caused by different environments of schools and different environments of managers. In the sense of objects, it refers to changes of ways and means in the management of managers due to diversity and variability of subordinates. During the compact stage of school culture, the way of the headmaster's leadership can be either management by walking around or thinking calmly and even might be passionate propelling and guiding effective guidance according to changes of circumstances, environments, objects and so on. Sometimes, the headmaster can be on the stage, and sometimes can be in the backstage or walk in the middle to lead. Such a trend in schools with weak culture characteristics is not obvious.

The second is that the cultural system of schools turns logic. During the stage of ripeness, "sides", namely the logic system of school culture, need to be refined and summarized consciously. It is advisable to adopt the narrative research approach. After years of development, the headmaster leads all the people in schools to think together: what did we do? In the school development history, who are key persons? Who are star teachers and students? What key events have schools experienced? How were they solved? How were the previous headmasters, teachers and students in schools? What information do architectures in schools convey? What are stories and ceremonies widely spread in schools? What is the relationship between these and the core values of schools? The headmaster encourages everybody to tell stories freely, find out stories with the highest repetitive rate, and make up them into a complete school story. Besides, the headmaster integrates the school stories and teachers' and students' feelings and narrations to research and extract vocabularies with the highest use frequency which can represent school spirit most, and each kind of spirit should have one classical school story as an explanation at least.

The third is the differentiation of school culture. After extracting "areas" of school culture, it is necessary to rethink about "points" of school culture, namely distinctive individual points and featured projects in school culture. These "points" may be systems, ceremonies or activities. It is necessary to let these "points" shine, widely spreading the school culture spirit; what's more, it is necessary to make these "points" become excellent classics and become symbols of school culture

Thus it is necessary to illustrate the relationship between school culture development stages and development states.

First and foremost, it is necessary to illustrate that the division and analysis of school culture development stages are carried out according to a rational linear fashion, that is to say, the way of school culture construction advances from linear and strap zones to continuous zone with constant speed. For a school, this state may exist during the time when a powerful headmaster continues working or when several powerful headmasters succeed successively. The state that the level of the leadership teams in schools is fluctuating or successors cannot carry forward is not discussed hereby.

Second, four stages of culture development in each school form a cultural development cycle, and this cycle will generally last for several years. After a cycle has been completed, school culture development will enter into the next cycle and go round and round. The several cycles will form a time spectrum of school culture development and these time features are the distinct reflection of social structure and its ideology in schools. However, this spectrum may not be a culture spectrum of linear development, and there may be ups and downs.

1.2 School Culture Structure

1.2.1 Introduction of the School Culture Structure

School culture structure means the consisting parts of school culture. Its composition varies when studied from different angles. And there are different opinions among scholars. At present, there have been four common divisions about the school culture structure, namely dichotomy, trichotomy, quad-chotomy and sex-chotomy.

1.2.1.1 Dichotomy

Dichotomy is one of the popular statements. One statement supports that school culture should be divided into the spiritual culture and material culture. Material culture is a surface structure while the spiritual culture is a deep structure. Another statement divides the school culture into system culture and non-system culture according to its state. In addition, according to the subculture group, it can also be divided into teacher culture and student culture.

1.2.1.2 Trichotomy

America scholar James L Heskett divides the structure of corporation culture into deep structure, middle structure and surface structure. Deep structure means the common values of corporation including corporation spirit, operating and managing ideas, professional ethics, attitude towards talent and sense of competition, which can be adopted as guidance by staff. Surface structure is the entity of material culture, such as technique and equipment, construction form, corporation environment, product design and operating style. Middle structure has something to do with system and behavior structure. Such culture structure is a pyramid existing in any organizations. Inspired by this, we can divide the school culture structure into deep structure, middle structure and surface structure, namely spiritual culture, system and behavior culture and material culture. This division is also very common and has many variants, such as material culture (surface), system culture (middle) and spiritual culture (deep) or entity part, entity and spirit part and spiritual part.[16]

1.2.1.3 Quad-chotomy

On the basis of trichotomy, quad-chotomy makes a further division and description. In such views, the school culture system is considered to consist

[16]Zhao Zhongjian, School Culture [M], Shanghai: East China Normal University Press, 2004: 117.

of spiritual culture, system culture, behavior culture and material culture. In the pyramid consisting of the four parts, spiritual culture is on the top; below it is the system culture, behavior culture and material culture. In some circumstances, system culture and behavior culture are mixed into one, called behavior culture. If so, the quad-chotomy is turned to be trichotomy and we can say they are almost the same.

There are also many variants of quad-chotomy, such as the division of material life culture, systematic administration culture, behavior and custom culture and spiritual consciousness culture, or the division of intelligent culture, spiritual culture, normative culture and material culture;[17] or the division of environment structure, system structure, behavior structure and spiritual structure. In addition, there is also penta-chotomy, which holds the idea that the school culture consists of core values, management culture, behavior culture, curriculum culture and material culture.[18]

1.2.1.4 Sex-chotomy

Wang Jihua holds that there should be six carriers in school culture: environment carrier, like campus design and landscape architecture; concept carrier, like school motto, school song, school badge, educational concept, educational aim and values aim; activity carrier, like school anniversary, memorial day, class meeting, flag-raising ceremony, art festival, sports meeting, interesting group and scientific and technological activities; instructional carrier, like class teaching of each subject; system carrier, like student rules and regulations, civilized pledge and management system; behavior carrier, like the spirituality of principal, teachers and students: school style, instruction style and style of study.[19]

To sum up, no matter it is dichotomy, trichotomy, quad-chotomy, penta-chotomy or sex-chotomy, there are no significant differences. According to the varying process of inside to outside and deep to surface, we can confirm that the consisting part of school culture pyramid are spirit culture, system culture, behavior culture and material culture. Such division is convenient for researching and implementing.

---

[17] Zhao Zhongjian, School Culture [M], Shanghai: East China Normal University Press, 2004: 118.

[18] Zhao Zhongjian, School Culture [M]. Shanghai: East China Normal University Press, 2004: 127.

[19] Wang Jihua. Executive Force of New Educational Culture [M]. Changsha: Yuelu Publishing House, 2008: 11.

1.2.2 Analysis of School Culture Structure

1.2.2.1 Spiritual culture of school

Spiritual culture of school is the manifestation of deep school culture. It is a kind of spiritual achievement and cultural conception created by staff or most of the staff through long-term educational practice, formed under the influence of certain social culture background and ideology. Spiritual culture of school is a value chain of school culture and also called educational conception system of school. Values are the resource of supporting power of spirit, which has the function of guidance and regulation for the behaviors of all members at school and provide a hand for survive and development. It's the core of school culture and includes core values of school culture, schools' cultivation objectives, schools' development objectives, school motto, school song, school logo and other elements.

1.2.2.1.1 *Core values of school culture*

Values are a systematic guidance with beliefs, tendency and attitudes, which decide the scope of judgment, choice and influence the decision and behavior. It can help the formation of organization culture and internal regulation. It's also the key to cultivate sense of identity which can combine the individual's goals and organization's objectives.

The core values of school culture are the educational philosophy of school with another name of formulation of school's mission and are also the essence of school culture. "It is abstracted from the diversified school values with basic information and can be chosen by individuals with different values."[20] It is the concise form of the whole complicated system of beliefs which can be expressed by phrases or sentences in slogan, for example, the core values of Zhongguancun Fourthly Primary School in Haidian district Beijing city (Abbreviated as Zhongguancun Fourthly Primary School) are three sentences: Everyone is important; everyone has its value; everyone can bring change. During the process of implementation of the core values, the school has obtained the spirit culture of research, autonomy and initiative. The core values of Firstly Primary School Fengtai district in Beijing city are to promote teachers' development with revolution of organization, promote students' development with teachers' development, and promote school's development with the joint development of teachers and students. During the process of implementation of the core values, school has established the

[20]Shi Zhongying. Discussion on School Core Values and the Formation [J]. K–12 Schools Management, 2008(10): 4–7.

spirit culture of truth, harmony and cooperation. After setting the core values, school can, according to it, set cultivation objectives, development objectives, school motto, school song and school logo. In other words, we can say all these elements are the expression and carrier of core values.

1.2.2.1.2 *Two objectives of school*

Two objectives of school are the cultivation objective and development objective. School's cultivation objective is also the educational objective and the answer to the question that what kind of talents the school wants their students to be; school's development objective is the goal of running a school and is also the answer to the question that what kind of school the school wants to be. Zhongguancun Thirdly Primary School in Haiding District Beijing (Abbreviated as Zhongguancun Thirdly Primary School) is a famous and comprehensive school in Beijing. Now, it's standing at the starting line. In 2012, it comes up with an idea of "Our School" and sets a new position for school development: being a common studying area with Beijing's specialty, world-class level and compatibility for everyone. Its educational objective is to cultivate students to be responsible, aggressive, rational and with the ability of logic and leadership.

1.2.2.1.3 *School motto and school song*

School motto means the phrases or famous remarks which are concise but have the function of guiding and inspiration. It's the expressing words of formulation of the school's mission and the motto of teacher and staff. It is not informative in contents, but should be obvious in culture characteristic. In the whole system of school's spirit culture, school motto is the most incisive and special with the highest spreading rate. Through school motto, we can directly obtain the educational philosophy, history and tradition, school specialty and the style and taste of the headmaster. The school motto of Jieyang Huaqiao High School in Guangdong Province is "Make self-examination and introspection on morality every day; Keep working hard on study" and the explanation for this motto is as follows:[21]

"Make self-examination and introspection on morality" comes from *The Book of Change·Da Xu·Tuan* which means obtain a new morality every day and contains two meanings, namely pursuing new and cultivating morality.

[21]Gao Huanxiang. Make self-examination and introspection on morality: School Culture Construction Plan of Jieyang Huaqiao High School [M]. Guangzhou: Jinan University Press, 2009: 33–36.

It means continuance of correction and is an addition to virtue. To set this phrase as school motto is to promote a kind of aggressive spirit which is the imitation of the universal movement. This is aimed at improving oneself, being aggressive, working hard and being someone. Pursuing ideal morality means continuing doing self-examination, doing morality cultivation, molding one's temperament, developing good characteristic, and being perfect. Choosing "everyday progress" as the creativity of "great virtue" which means destroying the old and establishing the new and making self-examination and correction with the view that not to advance is to go back.

"Keep working hard on study" comes from "Diligence makes excellent study, and indolence makes bad study; introspection makes good behavior, and arbitrariness makes bad behavior" in *Explanation for Study* of Han Yu. It means that diligence can make better study, but indolence can make it bad; twice thoughts can make success, but discipline will destroy the success." "Diligence" means try and continue to do more. For students, "keep working hard on study" means keeping studying every day to make it better, and in the aspect of teachers, it means keeping promoting one's professional ability and career development. Whether it is study or instruction, career or business, all of these need consistence, creativity and diligence to make continuous progress.

School song is one of the important carriers of school spirit culture. It can be used to show the history, tradition and geographical feature, to express the educational objectives and formation of school mission, and to narrate the common dream and pursuit of teachers and students. In a word, no matter what kind of formation or function is, it should be able to express the core values of school.

1.2.2.1.4 *School logo*

Each school shall have its own logo for it's the symbol of school and contains special meaning with magnificent characteristic. The design of logo can be divided into the concrete type and abstract type. Concrete type is to give specific things certain meaning and is used to show a kind of spirit and culture connotation. Abstract type is to express school's concept through association by using abstract image and symbol. For example, the school badge of Jieyang Huaqiao High School in Guangdong is inspired by Jinxian Gate. At the center of the picture is the building outline of Jinxian Gate, the three words and the picture of a book. The connotation of this school badge is recommendation, unification and desire for talents to raise the education level of Huaqiao High School and Jieyang city. "Expect talents to come" and "people who come

Figure 1.3 Logo of Guangdong Jieyang Huaqiao High School.

in are equal to talents. It propagates the educational concept of respecting talents and emphasizing on morality. That is to put individual cultivation in the first consideration of education and focus on emulation of good people, cultivation of virtue, "teaching students to be honest" and "learning to be honest". It is complementary with the school motto: "Make self-examination and introspection on morality every day; Keep working hard on study".

1.2.2.2 System culture of schools

System culture of school is the integrity of organization structure, rules and regulations and the management culture formed during the process of implementation of spiritual culture. System culture is one of the patterns of expressing and implementing forms of school's spiritual culture. It tells actors what should be done and what should be encouraged. Besides, it is beneficial to the spreading of the value system of school and regulation and supervision of discipline of teachers, students and staff.

1.2.2.2.1 *School's organization structure*

School's organization structure is the integrity of vertical working groups and horizontal departments and the relating relationship, including the organization structuring, position, power, communication and network and other regular and technique, internal organization, social psychology and other non-official parts.[22] The types of organization structure include linear type,

[22]Chen Xiaobin. Educational Administration [M]. Beijing: Beijing Normal University Publishing Group, 2005: 403.

functional type, linear-functional type, career type, committee type, matrix type and so on. The design of school's organization structure shall comply with the requirements of school's spiritual culture.

Zhongguancun Fourthly Primary School described its school organization structure in its 2009 School Culture Construction Manual as center + project team + individual studio. There are five centers, three projective teams and one individual studio in school. The five centers are as follows: Faculty Development Center which provides support for teachers to update knowledge structure in the teaching research, lift self-examination ability and practical ability to promote the formation of natural and unified educational concept of teaching group and form to be a joint good teaching group of Zhongguancun Fourthly Primary School; Student Development Center which creates a rich, colorful and charming school life for students, promotes the construction of a special and distinctive class, makes students develop into good habit and moderates behavior through planning activities, and gives equal practical chance to each students; Resources Information Center which provides teachers and students with abundant information, makes use of these information to conduct fast and careful service, and spreads school's culture through internet and other means; Quality Improvement Center which assesses and checks all the work, provides data information and facts support for improving management and teaching level, finds and spreads the spirit of excellent people and typical events that can show the culture of Zhongguancun Fourthly Primary School, and helps to perfect all the work of school; School Affairs Center which ensures the efficient operation of school and daily work. The project team is aimed at construction of teaching group. Different working groups make diversified organization constructions. Employment of project group has provided chance and room for self-directed development of different teachers. The construction of project group mainly focuses on promoting the development of teaching policy that can improve learning efficiency of students, creating a platform for teacher doing research and a culture that is supportive and open, and forming a teaching group that is flexibly available for different teaching task and goal. The three project groups are as follows: Interactive development project group is the group of heterogeneous teachers to develop teachers' social ability and cultivate team working sense; subject research group consists of teachers who are in the same or similar researching direction or project. They conduct the same teaching research and the contents are usually the same, such as the research on how to help students to develop the habit of listening; project recruitment group mainly focuses on the works in school time or temporary project, which is reported by teacher according to individual specialties and researching focus of current term. In addition, traditional project research group and activity group shall continue to implement the responsibility of coordination. Individual studio is set up typically for excellent head teachers. At present, there are two individual studios, namely Feng Chun Studio and Yang Lijun Studio. Head teachers who have come into trouble can turn to the two teachers.

The organization structure of Zhongguancun Fourthly Primary School is filled with new feeling and new meaning. The management idea of it is to "do things and educate people". The school holds that traditional career structure type and separated department structure are divided according to position and power, which has overviewed the function of people. People must be the core of organization. The organization structure of Zhongguancun Fourthly Primary School is a double structure, in which the center + project group + individual studio are the main structure and its aim is to finish all projects and work;

Such structure weakens the administrative power in vertical distribution and turns it into service and cooperation, but stresses teachers' function and considers students as centers. It truly shows the core values that "everyone is important, everyone has its value, and everyone can bring change".

1.2.2.2.2 *Management system of school*

Management system of school is a series of rules or regulations established during the implementation of education. For its constitutor, management system of school is divided into three kinds, namely national class, local class and school class. For managers, management system of school is divided into cadre management system, teacher management system and student management system. For functional department, management system of school consists of educational and teaching management system, teaching and researching management system, personnel management system, class management system and logistics management system. For projects and objectives, management system of school includes management system of moral education, teaching, physical education, aesthetic education, labor education and information and technology education. The formation of such system is that: responsible department prepare the draft—conduct survey and research wildly—research and discussion of key members—approval of Teacher's Representative Committee.

1.2.2.2.3 *Management concept of school*

Management concept of school is the expectation of school management. Humanism, personality, development, quality, efficiency, and discipline are usually the beliefs supporting the system. Management system is not rules but a comprehensive and detailed system connected by values. Management system has its own function—school system is based on core values of school, cultivation objective and development objective. System is not hanging for ornament but for providing guidance and standards. And humanized management is highly praised and advocated by more and more people.

1.2.2.3 Behavioural culture of schools

Behavioural culture of schools is the summation of the socially acceptable behaviors and manners and the activities serving as their carrier, which school members develop under the guidance of spiritual culture and the norm of systematic culture. It includes the behaviors and manners of the headmaster, the behaviors and manners of the teachers and the behaviors and manners of the students. Behavioural culture of schools is the activity culture formed during the process of educational practice; it can be subdivided into manners, activity and ceremony of people, featured projects of school and so on. Behavioural culture is the "second foot" of school and is the important manifestation way of school culture, which demonstrates the characteristic style of school as well as reflects the spirit of school.

Behaviors and manners are the etiquette and norm observed by people in their speech and deportment, manners of dealing with people and communications. The behaviors and manners of school mainly include the headmaster-centric behaviors and manners of administrators, teachers and students. We can observe and study the behavioural culture of schools by combining the above mentioned behaviors and manners with some appropriate activities or ceremonies.

1.2.2.3.1 *Behaviors and manners of administrators*

Headmaster culture is the soul of school development and the unique spiritual pursuit and behavioral features of headmaster reflected from the cultural process of school.[23] The behaviors and manners of headmaster mainly refer to the way and style of the leadership including the ways of self-leadership and the way to lead others and groups. The construction of the behaviors of

[23] Wang Jihua. Executive Force of New Educational Culture [M]. Changsha: Yuelu Publishing House, 2008: 11.

the headmaster mainly refers to the improvement of the leadership while the behaviors and manners of the middle-level cadres of school mainly refer to the way and style of their management including manage others and departments, as well as self-management. The construction of their behaviors mainly refers to the improvement of their management and execution ability. The foresaid two types of initiative person constitute the management team of school. Their behaviors and manners are associated with the training objective and development goal of school; moreover, the implementation of their behaviors and manners are combined with the featured projects of school. The culture can be manifested by the strong project or traditional features of school, for example, Kongqiyangren culture of Beijing Second Middle School and the Beat Dates Festival during the autumn of Beijing Chaoyang District Hucheng Middle School, these are ceremony and activity excavated and designed on the basis of the school culture; the activity "Be young Gentleman And Read Classic Works" of The Primary School Attached To Peking University is of great momentum, and it is bound up with the taste, preference and behaviors and manners of the headmaster management team.

1.2.2.3.2 *Behaviors and manners of teachers*

Teacher culture is the summation of the behaviors and manners of teachers. The behaviors and manners of teachers mainly refer to behavior norms and activity structured around their professional development including teaching behaviors, study behaviors, courses culture and so on. It relates closely to the objective, management system, management style, management activity and other aspects of school. The culture of study and courses culture are the key part of teacher culture. Lesson study is an important cultural tool to improve the professional capability of teachers, and cultivate the culture of study and instruction culture.[24]

1.2.2.3.3 *Behaviors and manners of students*

Behaviors and manners of students refer to the reflection on behaviors and manners of values and cultural orientation in study, life and work owned by all students within the specific social space.[25] Behaviors and manners of students mainly refer to the behaviors norms and activities and ceremonies and so on

[24]Zhang Dongjiao. Lesson Study: Cultural Tools to Improve Schools' Efficiency [J]. Journal of The Chinese Society of Education, 2009(10): 21–25.

[25]Wang Jihua. Executive Force of New Educational Culture [M]. Changsha: Yuelu Publishing House, 2008: 13.

structured around healthy growth and development of students. They can be implemented in aspects such as class culture, model students, featured activity and ceremony, story and legend. Class culture is the approved atmosphere with sense of belonging, culture type and value orientation collectively built by the head teacher, subject teachers and all students of the class.[26] For example, the "phenomenon of writer groups" of Beijing Second Middle School takes advantage of the appeal and stimulation formed by successful model graduate groups. The portraits and masterworks of dozens of famous writers graduated from Beijing Second Middle School are demonstrated in the showcase; it becomes a long lasting legend. Beijing Fengtai Caoqiao Primary School has carried out "Gratitude Education" since 2005 by bringing up "To Start the Gratitude Education by Being Grateful to Parents". The activity covers the gratitude to classmates, teaching staff, society and other aspects later, and many relevant special activities are organized effectively. Students' activity with unique features of Caoqiao Primary School has been formed during the process. For instance, let students experience the hardship of their parents through "carrying the schoolbags upside down" and family work and other activities; to cultivate their initiatives to learn and develop the good habits of not littering around or wasting food and being polite to teaching staff through experiencing the work of teachers, cleaners and workers in the canteen, as well as carrying out investigation and observation. Through the gratitude education, students in Caoqiao Primary School behave politely and become readily to help others in the community. As a result, the school often receives commendatory letters from the community and parents.[27]

1.2.2.4 Material culture of schools

Material culture of schools is the external symbol of school culture and the school level culture expressed in the form of material created by the school member during the educational practice. It includes the school architecture, campus landscape, cultural facilities and other aspects. Material culture of schools is the "third foot" of school culture and always related to audio-visual recognition system, and it is the materialized expression form of school culture which visualizes the school culture and spirit.

[26]Wang Jihua. Executive Force of New Educational Culture [M]. Changsha: Yuelu Publishing House, 2008: 14.

[27]Gao Yimin. Assessment Report on the Cultural Development State of Fengtai Caoqiao Primary School [R]. 2010: 1

1.2.2.4.1 *Architectural culture of schools*

Architectural culture of schools includes the designs and carriers of the style of school architectures, basic colors, buildings and roads, offices, classrooms, corridors and other culture elements.

The development history of school architecture can be divided into three stages: school architecture without specific image (it refers to the school architectures during the period when schools did not exist), school architecture with distinct characteristics (it refers to the school architectures during the period when schools exist) and school architecture which doesn't look like school. By far, we have entered the last stage which can be called the stage of people-oriented new type school architectures. Its style and features are: multifunctional open-end spaces replace the close-end spaces connected by corridors; the schools are transformed from education-oriented spaces to study-oriented spaces; importance is attached to the influence of indoor and outdoor environment and space atmospheres on students' physical and psychological health and sentiment; diversification of shapes, colors and space forms; the open and integration of school to society and community. The change of styles of architecture is the expression form of the profound change of educational idea, thoughts and educational approaches.[28] For example, the designs of architecture of Beijing Haidian District Second Experimental Primary School are distinctive nationwide: colors and geometric figures are used in campuses such as the color match of light pink and gray on the buildings, classic color match of blue and red in the hallway, which are bold and of visual impact; there are rounded windows on classrooms' doors; the safe wall in main campus; safe pillar in Qingning Campus; graduate bottles in Dangdai Campus. All of them are distinctive.

1.2.2.4.2 *Campus landscape*

Campus landscape is one of the material carriers of school culture, and is the interpretation and expression of the core values of school. Campus landscape includes natural landscape and human landscape. The natural landscape refers to the natural ancient trees and rare stones in the campus while the human landscape is artificial. Inscriptions of famous people and campus sceneries are important human landscapes. The natural landscape and human landscape usually integrate with each other. For example, the main building of the Primary School Attached to Peking University is built near trees, known

---

[28]Zhao Zhongjian. School Culture [M]. Shanghai: East China Normal University Press, 2004: 85.

as Ecological Building, which is quite distinctive. There is the Ecological Building embraced by ancient trees and the twelve scenes of human landscape, inscription of Bingxin: learn intently and play heartily as well as the former home of Jian Bozan, and the campus itself is the garden of Wang Shixiang... The landscape itself is the abundant social capital of the school. These are witnesses of the history of the school; they reflect the length, width and height of the school culture.

1.2.2.4.3 *Cultural facilities of schools*

Here, cultural facilities of schools mainly refer to library and network. Library is an important site for expressing the school culture. School without books can't be called as school. There should be various types of books in a school library and a library should be designed from the aspect of culture. The Primary School Attached to Capital Normal University works hard to make reading become a culture action between teachers and students and takes full use of limited resources of books by not taking books as the decorations on the shelf; instead, school encourages students to read books in turn, and exchange books with other students in other classes, thereby forming the atmosphere of motivating oneself by reading and information sharing. It makes students to be accompanied by books and enriches their connotation and thereby become the subjects to promote the reading habbits. Teacher library motto "Erudite and Ambitious, High Virtues and Aspirations" and students library motto "Read and Establish Virtues, Learn and Become Talent", as well as the great upsurge driven by them, are motivating the advancement of reading in the campus.

1.2.3 Cultural System of Schools

1.2.3.1 Fishbone diagram of cultural system of schools

Cultural system of schools consists of spiritual culture, systematic culture, behavioural culture and material culture. We use Figure 1.4 to demonstrate the system.

1.2.3.2 Tripod-shape theory for school culture

The structure of school culture can be demonstrated by the pyramid-shaped figure in Figure 1.5.

Apart from the pyramid-shaped structure, the study holds the point that the school culture should be a situation of tripartite confrontation; brings forth tripod-shaped theory for school culture, namely, the spiritual culture is the

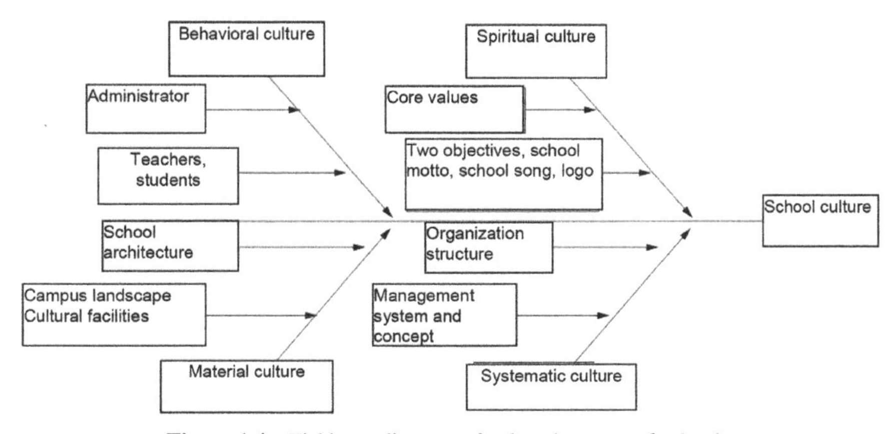

Figure 1.4 Fishbone diagram of cultural system of schools.

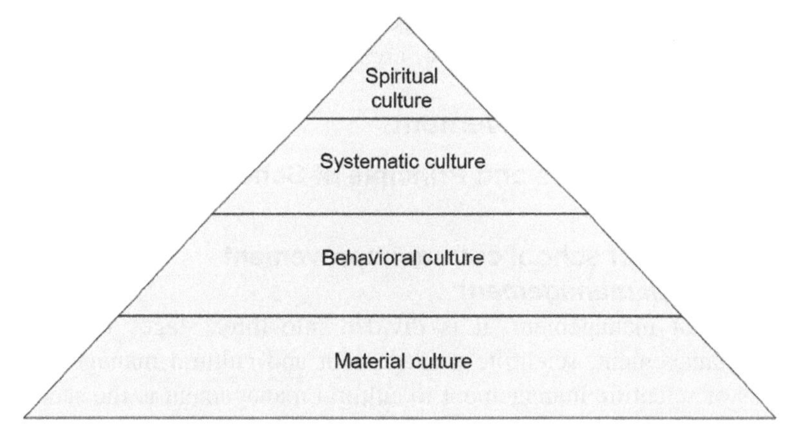

Figure 1.5 Pyramid shaped structure of school culture.

core and soul of school culture; it is equal to the brain of human and the body of tripod. While the systematic culture, behavioral culture and material culture are the reflection, carrier and expression form of the spiritual culture of school; they are equal to the limbs and body of human or the three legs of tripod. This is the "Tripod of School Culture"—culture is the tripod of establishment of school to school development.

1.2.3.3 Characteristics of cultural system of schools

School culture is something you can't buy or take away or remove. It is irreplaceable. School culture is the culture bred and soaked in school instead

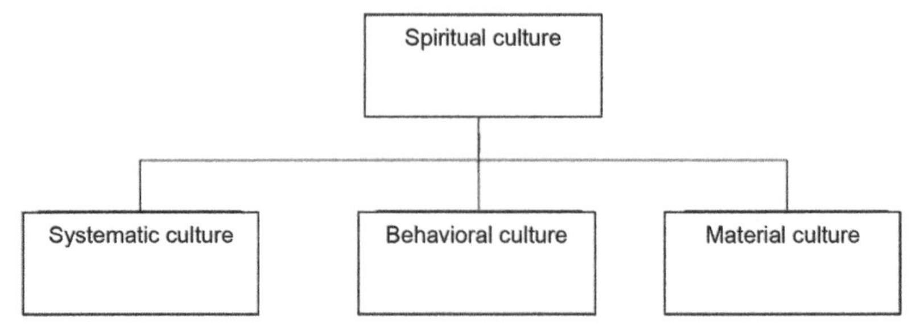

Figure 1.6 Model of tripod structure of school culture.

of foreign culture; it is actual and scene culture instead of conceptual or static culture; it is valuable and all-round and realistic and feasible culture. School culture is a lively culture created by all members in school one step by another. School culture is pervasive. School management is the management of school culture and each member is the initiative creator of school culture.

1.3 School Culture Improvement

1.3.1 Definition, Features and Principle of School Culture Improvement

1.3.1.1 Definition of school culture improvement

1.3.1.1.1 *Cultural management*

In the history of management, it is divided into three stages including experience management, scientific management and cultural management. The process of scientific management to cultural management is the second leap in the history of management.[29] Cultural management firstly appears in *Corporate Cultures: the Rites and Rituals of Corporate Life* written by two American managerialists Terrence E. Deal and Allan A. Kennedy. The concept of enterprise culture management is formed on the basis of enterprise culture concept and practice. And then, cultural management is separated from enterprise culture management and becomes a management idea, theory and method to all organizations. And schools are in no exception.

Frederick Winslow Taylor, the father of scientific management, points out that scientific management doesn't manage matters about science. So cultural management doesn't manage matters about culture, either. Three meanings in cultural management are: first, cultural management, an idea

[29]Zhang De, Wu jianping. *Cultural Management: Transcendence to Scientific Management* [M]. Beijing: Tsinghua University Press, 2008: 20.

and current of management, regards values as the core to form a school of cultural management. Second, cultural management, a management theory and a new stage of management theory development, is a management thought and idea taking people as foundation. Third, cultural management regards organizational culture construction as management modes of central works to management, including methods of management practice applied by modern organizations through cultural management idea, doctrine and theory.[30]

1.3.1.1.2 *School culture improvement*

School culture improvement is not only constructive design and operation for four specific variables of the school, but also a process of trying to build culture-oriented school on the basis of former situation. School culture improvement, first of all, is aimed at the management of values, and then the management of systems, behaviors and materials. School culture improvement is a process of establishment and cultivation, which is full of several structures, factors and levels and also cultivates school values and constructs a community of study for all teachers and students. To reach the final goal of developing the school and educating people, school culture leadership and improvement are usually adopted. On the contrary, it's also the best way to build school culture

1.3.1.2 The feature of school culture improvement

1.3.1.2.1 *Hypothesis of people-oriented management*

The center of management is the basic sign to distinguish management theory and pattern. Cultural management is a people-oriented management idea and pattern. So, people are the core of management instead of the edge, the aim of management instead of the method. Cultural management emphasizes on human nature but never denies physical property. Human nature will not exist without physical property.

Hypothesis of cultural man is put forward in cultural management written by Terrence E. Deal and Allan A. Kennedy. In the hypothesis of cultural management, the relationship between the environment and people is that the environment is independent variable and people are dependent variables. In addition, the future of people is unpredictable. Zhang De puts forward the hypothesis of people with cultural management concept. In this hypothesis, people with this concept are complex, variable, subjective and initiative. This hypothesis emphasizes decisive influence of world outlook, outlook on life and

[30]The same with ①, page 28 and 29.

values and other basic concepts on human behaviors. Through affecting and changing these concepts, human behaviors will be influenced and changed enduringly, especially in the respect of ideal, belief, values and morality. The core of cultural management is to build common value or to implement management based on value.[31] People who are at the stage of development are the core of school education. So, the hypothesis of people with concepts is suitable for the school.

1.3.1.2.2 *School culture is the core job of management*

Now, Terrence E. Deal and Allan A. Kennedy wrote: When we wrote *Corporate Cultures: the Rites and Rituals of Corporate Life* 20 years ago, we never thought that enterprise culture at present would be an obvious management fact. All concepts are expected to be understood by contemporary managers and to be applied into the daily management work. But who would think in this way 20 years ago? We devote ourselves to seizing and explaining internal operating methods with deeper symbolized meaning in modern organizations. Nature of the mind is hidden under the reasonable appearance, which can't be changed by external technologies or subjective wishes. People who are in the culture field get happiness and benefits while ignoring the risk of chaos and crisis.[32]

The principal contradiction settled by cultural management is between comprehensive development and organizational development. To regard development of people and the school as the goal of school management in perfect harmony, it's necessary to regard construction of school organizational culture as the core work of organizational management. Cultural management is a management choice in the mature period of school organizational development. Cultural Management takes people as foundation with care, harmony, unity and other interpersonal factors to a great extent, which is the core work of school development.

1.3.1.3 The principle of school culture improvement

The principle of school culture improvement is to disclose and reflect movement rule of principal contradiction in school culture improvement and is the concentrated reflection of school culture improvement

---

[31]Zhang De, Wu Jianping. Cultural Management: Transcendence to Scientific Management [M]. Beijing: Tsinghua University Press, 2008: 62.

[32]Terrence E. Deal, Allan A. Kennedy. Corporate Cultures: the Rites and Rituals of Corporate Life [M] translated by Li Yuan and Sun Jianmin. Beijing: China Renmin University Press, 2008, Preface.

1.3.1.3.1 *Leadership and management*

The leadership is a kind of action or behavior, a power relation, and a method to achieve one's goal. Various definitions of the leadership have common elements that the leadership is a process with influence and appears in a group environment with a goal. Management is an activity in which some tasks and goals will be realized and organized by management official leaders. There are differences between leadership and management. Warren Bennis and Burt Nanus thought that management was to do something correctly while leadership was to do something right. Kurt pointed out that an overriding task of management was to ensure consistency between organizations and orders while main function of leadership was to produce changes and movements. Management and leadership respectively pursue orders and stabilities, and adaptive and constructive changes.

Leaders of school culture grasp correct and advanced nature of construction direction in school culture and encourage all staff to work hard. School culture improvement is a process to ensure effective realization and practice of every measure and step. It's necessary for school culture construction to combine a long-term view with earnestness and combine strategies with tactics.

1.3.1.3.2 *Thoughts and actions*

In the meaning of implementation, practice is the first rather than knowledge. Members of the school are pragmatic and agile actors who pay much attention to practice after having a promise, claim to take actions timely. School culture improvement is a pressing issue which should be considered in practice, combined perfectly with systemic thoughts of more powerful actions. Actions without goals are blind, while ideals without actions are empty. In the process of school culture construction, theory will naturally be coincided with practice.

1.3.1.3.3 *Facts and concepts*

School culture is the internal culture of facts and contexts, instead of pure concept culture constructed directly from top to bottom. So managers need to seek culture from historical facts and contexts of school organizations. In other words, the starting point of school culture construction is the culture of facts instead of various definitions of culture. In the process of constructing school culture, every school should directly face the history and facts of the school, understand cultural connotations of organized activities and historical facts to seek or construct spiritual motive in order to drive organized activities. In the mature development stage of school culture, culture can be summarized and refined in accordance with proper concepts to provide related words for expressing culture and to construct cultural system and logic, and culture can be perfected and amended gradually.

1.3.2 Main Bodies of School Culture Improvement

Main bodies of school culture improvement include headmaster, staff and students. They are all active creators of school culture. School culture improvement requires their co-participation and needs them to perform their respective duties, and work together to contribute their wisdom and strength.

1.3.2.1 Headmaster is a symbolic manager[33]

In an organization where culture has continuous features, managers take the lead in supporting and shaping culture. These managers are, of course, symbolic managers in that not only do they spend much time in thinking about value, heroic figures and ritual ceremonies about culture, but also believe that their ground work is to deal with the value conflicts occurring in daily work.[34] In order to carry out cultural management, headmaster has to become a symbolic manager.

There are remarkable differences between symbolic manager and non-symbolic manager. Headmaster, as a symbolic manager, is much sensitive to school culture and the importance of long-term success. He has enough insight and persuasion in criticizing and inheriting original culture or conducing new exposition, and is able to constantly put forward new culture references and frameworks. He highly trusts his staff and depends on these partners to head for success. He guides them through encouraging people around to take more efforts and inspiring their innovation spirit. In culture protection, he possesses high self-consciousness and responsibility, and is willing to grant his power to others and work with others. He grasps every opportunity to strengthen, magnify or permeate cultural core values. Headmaster is responsible to make us live in culture. Through the promotion

---

[33]Those managers who have stong enterprise culture consciousness and shape and support enterprise culture actively are called, by Terrence E. Deal and Allan A. Kennedy, as "symbolic managers". The reason why headmasters are called as symbolic managers is that they value the importance of culture a lot in management. They insist in a common value concept and make it permeate staff's ideology so as to be an organic component of their personal value system.

[34]Terrence E. Deal and Allan A. Kennedy. *Corporate Culture: The Rites and Rituals of Corporate Life*. [M]. Translated by Li Yuan and Sun Jianmin. Beijing: China Renmin University Press. 2008: 140–142.

from culture to school so as to make changes, the headmaster builds an environment suitable for human's development. The main purpose of school culture improvement lies in, through transferring school-running idea and philosophy into teachers' common pursuit, promoting headmaster's change from a school manager focusing on specific matters to a thinker focusing on school culture.

1.3.2.2 Teacher is a culture creator

Function of teachers is embodied in teacher culture construction. Teacher culture is combined by value system and behavior standard in school teacher community. It includes teachers' belief, value, attitude, habit and behavior standard.

Teacher is one of the construction main bodies and management main bodies, playing an ongoing positive role or a negative role in the course of school culture construction. Teacher's positive attitude to school culture appears in the aspects of belief, attitude and behavior standard. The direction and degree of its function depend on how a teacher gives play to his authority and construction of teacher-student relationship and teacher-teacher relationship. Teacher culture construction requires beginning with inter-personal culture management.

1.3.2.2.1 *Being good at managing teacher-student relationship*

Teacher, as the manager of teacher-student relationship, has to be a person good at communicating with students and learn to be a democratic teacher-student relationship manager. A research conducted by Ronald Lippitt and Ralph K. White showed that teacher's leadership and his behaviors and manners were important for class atmosphere and teacher-student interaction model. They raised four types of teacher-student relationship and learning behavior resulting from this relationship.[35]

Tough and arbitrary type: Teacher often supervises students severely. He thinks that praise could spoil children, as students cannot study without teachers' supervision. Students' reflection: they yield to but disgust this kind of leadership, tending to shirk their responsibility, easy to be angry and unwilling to cooperate. As soon as teacher leaves classroom, students are absent-minded.

Merciful and arbitrary type: Teacher does not consider himself as an arbitrary person. He praises and cares about students, defining "ego" as a

[35]Quoted in: Zheng Jinzhou. Education and Culture Studies. [M]. Beijing: People's Education Press. 2000: 274–275.

standard of all class work. Most students like him and depend on him, resulting in the lack of creativity of students. Class work is of high quality and much quantity.

Laissez-fare type: Teacher lacks confidence in getting along with students, or thinks that students can do whatever they want. He is hard to make a decision, and has no clear objective. Neither does he encourage students nor rejects them, and neither takes part in students activities nor offers help and methods. Students are poor in both morality and study. There are behaviors like shirking responsibility, seeking for a scapegoat or easy to be irrigated. No cooperation. No one knows what he should do.

Democratic type: The whole class makes plans and makes a decision together. In the premise of no harm to collectivity, teacher is glad to give individual students help or guidance. He encourages collective activities, and gives objective praise and criticism. Students are fond of learning, and fond of working with others especially with teacher. Student work is of high quality and quantity. They encourage each other, and shoulder responsibilities independently.

1.3.2.2.2 *Being good at partnership*

Teacher is also a teacher-teacher relationship maintainer. Teacher culture is reflected by features of relationship and communication between teachers. Andy Hargreaves divides teacher culture into 4 types.[36]

Discrete culture: Most teachers work in their independent ways, form a series of separate kingdoms, and are unwilling to cooperate or interact with others.

Balkan culture: A school is divided into more than one independent or even competitive team where teachers are loyal to their own factions. There is no communication or concern between factions. With the interest conflict worsening, frictions may occur between factions. They even try their best to get maximum benefit from the interest of others.

Natural cooperation culture: Not only do teachers accept others' observation publicly, but also observe others' courses. And teachers discuss their experience in observing courses, and are willing to try the reform.

Factitious culture: Culture atmosphere between teachers is formed in the way of increasing opportunities of teachers' joint programs and mutual

---

[36]Zhao Zhongjian. *School Culture.* [M]. Shanghai: East China National University press. 2004: 396.

learning through a series of normal, specific bureaucratic procedures including forced team teaching, providing conditions for cooperating plans, and arranging instructors for new teachers. This culture encourages connections between teachers, professional experience sharing, learning and improving. And it also helps the materialization of new methods and new techniques.

In developing new professionalism, teachers advocate transferring form autonomous individual development focusing on teachers to peer interaction and cooperation culture focusing on teachers' professional development. They lead to establish positive partnership, and multi-direction learning. And they strengthen professional talk, communication and cooperation. In the early 20th century, Niels Henrik David Bohr, a famous Denmark scientist and the leader of Copenhagen School, positively encouraged a strong academic atmosphere characterized by equal, free discussion and close mutual cooperation, namely, Copenhagen spirit, in Weil Institute Copenhagen University. It is a teamwork principle, and a team built depending on this principle is one with Copenhagen environment.[37] Leader allows teachers to involve and accept school vision, to understand their own roles in this vision, and to make contributions in cooperation in executing the vision. Headmaster leads the team to face the power sharing and to propel the school development so as to make teachers feel the strong support from internal school. It encourages teachers to shoulder leadership responsibility, and respects their autonomy.

1.3.2.3 Student is a culture displayer

Student should keep tension and elasticity among leader culture, teacher culture and student culture, forming a main body same features among these three culture main bodies. Culture elasticity is an important index testing culture activity. Whether school culture is alive?[38] Degree of students' involvement to school culture is one of the important indexes.

Student is also one of the important creators of school culture whose function is embodied in student culture construction and management. Meanwhile, student is the final displayer of school culture. Managing student culture requires initiating students' self-management as well as cultivation first.

[37]Xia Zhongyi. *University Humanities Reading: Man and the Word.* [M]. Guilin: Guangxi Normal University press. 2002: 209–212.

[38]Tang Jinyin and Sun Jialin. *Study of School Culture: Perspective to a Middle School's School Culture* [M]. Beijing: Education Science Press. 2005: 143.

1.3.2.3.1 *Setting good examples*

As culture without model and hero is imperfect, bring model into playing the effect is an important way for constructing and managing student culture. Set those students who can best reflect school value and spirit as models, publicize and praise them, which is beneficial for formation and growth of outstanding school culture, and has function of leading, collecting and coordinating.

School should be good at finding original student models, nurturing and encouraging them by adapting various methods such as symbolic behavior – setting headmaster award, school ritual – raising a flag, commencement and opening ceremony, and student story – famous or heroic schoolmates' stories handing down. School should not let any publicizing opportunity pass, as every opportunity is important rather than no worth mentioning.

1.3.2.3.2 *Self-management*

Student is also one of the school culture managers. Student autonomy can reflect cultural management spirit, and self-management is an ideal goal education pursues. Student plays a very important role on the stage of self-management in class, grade and school. In our country's middles and primary schools, student cadres have been adopting a management system in which political system and administrative system in parallel. Student cadres are composed of two parts: one is political system consisting of class Youth League (Young Pioneers) branch committees, and usually sets League branch secretary, commissary in charge of organization and commissary in charge of publicity; the other is administrative system consisting of more than one class committee, and sets monitor, commissary in charge of study, commissary in charge of sports, commissary in charge of literature and arts, and commissary in charge of general affairs. Class student cadres are composed of League committees and class committees (called "two committees" for short) with 8 members. In order to guide students' self-management, Beijing Chen jinglun Middle School with 89 years' history has carried out the improvement of student cadre management system nationwide in the form of systematization, gradually forming a kind of management model called Chen Jinglun middle school student cadre "ministers and commissions system" which is a new and timely attempt.[39]

---

[39]This case is provided by Zhu Hongqiu, deputy headmaster of Chen Jinglun Middle School in Chaoyang District of Beijing. Now extend our thanks for him.

Chen Jinglun Middle School's school-running guiding ideology is "constructing personalized school, attaining personalized teachers and nurturing personalized students". And its education goal is "enabling every student with health, knowledge and reason fly from Chen Jinglun Middle School, soaring in the sky of twenty-first century". To achieve this goal, the school makes a reform for class student cadre establishment. 16 student cadre management positions are set in every class, which increases cadres including commissaries in charge of science and technology, hygiene, public property, information, electrified education, books, psychology respectively, and Love president, and increases the number from 8 to 16. To effectively join the management of student cadres in school, grade and class, the school has established Chen Jinglun Middle School Students' Self-management Committee (called SMC for short). SMC sets 16 "departments" including secretary's work department, organizational work department, publicity work department, class work department, work departments of study, sports, literature and arts, general affairs, science and technology, hygiene, public property, information, electrified education, book, psychology respectively, and Love work department.

Every "department" is composed of corresponding student cadres in every class. And every "department" sets one minister and four vice-ministers in charge of junior, senior 1, senior 2 and senior 3, and professionally manages corresponding work in school, grade and class. The school joints all "departments" to corresponding offices and professional teachers according to its work requirements. 16 "departments" are attached to 4 offices including School League Committee, Moral Education Department, Department of Education, and Property Management Office, and according to their administration authority joint with related offices including sections of sports, arts, information, electrified education, maintenance respectively, library, psychological room, and clinic. From then on, all professional work is mainly carried out by offices through corresponding "departments" instead of through Moral Education Department – head teacher – student cadres. As a result, information transmission efficiency and work management efficiency have been increased a lot. Meanwhile, the purpose of promoting overall moral education and lightening head teacher's overmuch burden has been achieved. At last, culture value that encouraging students' personality has been inherited.

1.3.3 School Culture Improvement Strategy

School culture improvement strategy focuses on spiritual culture, systematic culture, behavioral culture and material culture and is mainly to conduct and promote the school culture orderly and efficiently. It contains four strategies, namely enlightening strategy, talent-use strategy, affairs management strategy and materialization strategy.

1.3.3.1 Enlightening strategy

Values management is the core of school culture improvement and aimed at developing common values. Enlightening strategy is for spiritual culture management of school. Spiritual culture management is to summarized and explore philosophy of school education based on completely mature culture facts, and to systematize conception of school as well as describe school culture as tabbed conception. A school with strong culture must possess a relevant stable core value. Under this premise, the value system of school can make adjustment and new statement according to environment change to demonstrate the change of time and the improvement of knowledge. Then how to make use of enlightening strategy?

1.3.3.1.1 *Explore and refine educational philosophy from facts*

School culture improvement is a long-term and tough project, and refinement of culture is also not a relaxing procedure, which needs time and wisdom. Refiner may be the member of a school or a joint working team consisting of invited educational experts and school staff. Refiner must respect the cultural facts of school; know about the history of school and all statements relating to school culture; read all the typical stories about school; find out original experience and tradition by historical narration method; encourage all the school staff to participate in this activity, and hear them talk about all existing school conceptions, stories and legends of past and present; use research method to help cadres, teachers and students to know the key words that can express the school spirit; judge whether the school belongs to school with weak cultures or strong cultures; hold a brainstorm to gather inspiration; conclude and refine the core value of school and at appropriate time revise and perfect it; bring up one or several readable words or sentences that can express the essence of the core value.

1.3.3.1.2 *Turn the system of school spiritual culture to be logicalization*

After finding the core clue of educational philosophy of school, make it logically fluent, reasonable and coincident according to spiritual culture system or value system of the organizational school. School's cultivation objectives, school's development objectives, school motto and logo design are the presentation of core values and the important elements of value system.

1.3.3.1.3 *Refinement of the school spirit*

On the basis of getting comprehensively familiar with development status of school culture, sum up school's spirit that is coincident with educational philosophy of school and conform to the cultural facts of school. Use one or several accurate words or sentences to make it concise, of which the difficulty is that it shall be coincident with experience but also be superior to experience. The six-little-citizen education system of Beijing Fengtai First Primary School is a very mature and special example. The school's spirits are "Pragmatic, Harmony and Cooperation". For nearly half a century, the school constantly put their working focus on the core values of "Pragmatic", which means "dealing with concrete issues", and apply it to practical workings truly, attentively and steadily: focus on fundamental work, stress practical efficiency and have clear thoughts and ideas. This school regards instructional quality as foundation for forty years. Stressing practical efficiency means that managing group and leaders of school must manage to realize such change and efficiency: students can grow healthily and teachers' professional ability can be developed steadily. Clear thoughts and ideas mean that school working shall be done step by step carefully and attentively based on certain guiding thoughts, guiding strategies and executing methods. Interview, lecture and cooperation are the basic methods of school for conducting feedback, thinking and studying. Teaching is the base, researching is root, and curriculum is the branches. The development of school is the representation of the realizing of this value. Such spirit is reflected from the construction of school's system and behavior system, also reflected from the interpersonal culture and task culture of school. Harmony is an interpersonal culture stressing harmony. Fengtai First Primary School is excelsior, pragmatic and emphasizing efficiency. It aims to be harmonious on personality and seeks to make progresses jointly. There is personality equality between teachers and students. They study from each other, and help each other. Teachers cooperate with each other harmoniously, but they are harmony in diversity. The construction of school's interpersonal culture is conducted though centering on the core of "three conversations". "Three conversations" mean communicating with students to prepare learning-based instruction, and constructing instruction and learning; communicating with peers to do research jointly and improve the quality of education; communicating with oneself to conduct analysis and introspection. Cooperation is a kind of task culture aiming at reaching a state of cooperation. Task culture and interpersonal culture are the expression of objectivity on institution and behavior. The construction of task culture of Fengtai First Primary School is conducted from inter cooperation of school, outer cooperation of school and interschool cooperation based on the principal of cooperation. The inter cooperation of school includes the combination of teaching and researching and three-site cooperation. Outer cooperation of school includes cooperation with parents, community and army. Interschool cooperation includes cooperation among schools including cooperation of college and primary and primaries.[40]

1.3.3.1.4 *Obtain proper conception*

School's culture is a living fact culture. A school with strong culture in a proper time may consider adopting a conception that can express school's culture. For example, Primary School Attached to Capital Normal University makes a summary of its school's culture as education for childishness that is to create an exploring fairyland, build a love school, and light a candle for children. Beijing Fengtai Fifth Primary School defines education for happiness as inspiriting ourselves, being caring, creating slenderness and accumulating happiness. The Peixing Education of Beijing Haiding Peixing Primary School and development education of Chinese Academy of Agricultural Science Elementary School both are the successful models. However, if the state is not so mature, this process will be unnecessary, because conception is just a form of expression, and culture is more important than it.

1.3.3.2 Talent-use strategy

Talent-use strategy is a behavioral culture with the center of promoting the development of teachers and students. Talent-use is not only the procedure of cultivating the culture of teachers and students but also the procedure of implementing and accumulating of school's core. For the principal and the whole leader group, this is full of challenge and is the key point for working creatively. Generally speaking, there are two kinds of managing strategies.

1.3.3.2.1 *First is the classified teacher cultivation strategy*

Classified teacher cultivation strategy is to make a classification according to the working experience and development degree of their profession and

[40]The case material is from Beijing Fengtai First Primary School. Thank Principal Gao Xiuxian for his efforts.

then conduct the cultivating working. For different schools, the number of teachers, teaching experience and the development of profession are different. Teacher needs to be "cultivated". During the management of school's culture, planning of management and its implementation, school needs to make a guiding supporting plan for teacher's profession development consisting of cultivation by new teacher, cultivation by experienced teacher, cultivation by famous teacher and cultivation by special-class teacher and other subsidiary plans.

The cultivation of teacher may use following contents and forms for reference: conduct the training of school's culture in groups or in the form of lecturing by one teacher; in the form of experienced one training new one; communication skill training; launch teaching skill training; make a group for new teachers; cast some salons for new teachers and so on.

The cultivation of experienced teacher is very important. A new teacher after several years' working may be qualified to be an experienced teacher. However, a teacher who even has an experience of eight years or more, if he is not excellent in his project, may be stuck in choke point and his training and cultivation may face bottlenecks. Thc cultivation for experienced teachers may use following contents and forms for reference: encourage teacher to further study to gain educational master degree; learn the scientific guidance on research and break the barrier of experienced research; study, read and enrich the knowledge structure; stimulate the PCK[41] development of teachers by using of lesson study. From April 9, 2008 to July 8, 2008, the project group in expert team of Beijing Normal University cast three lesson studies at Beijing Fengtai First Primary School, Bejing Fengtai Fifth Primary School and Beijing Fengtai Experimental Primary School. The three lesson studies are: promoting the development of new teachers: lesson study of *Chicken And Rabbit in One Cage* which is a mathematic question in grade four of primary; the lesson study is a practice study aiming at improving the oral skill of grade one through reading: lesson study of *Changeable Pupils* which is a passage in grade one; composition instructing class to let students "like to express, and love to write": "I am a__________ child", which is a Chinese passage study of grade four. Two teachers who conducted the lesson study are the Chinese teachers with 10 years' experience. The lesson study has improved the teacher's PCK and settled the problem of what to teach, how to teach and whom to teach. After the second round, the teacher used *Chicken and Rabbit*

[41]PCK, short for Pedagogical Content Knowledge.

in One Cage makes an investigation of learning situation and found that the number of students who have taken Olympic class after class is less than one third; students who can make the calculation do not understand the relationship of "counterpart difference". According to the result, the teacher changes the teaching objects and starts to change the strategy of combining the equality calculation, drawing and listing. After being changed to a question with small number, this question can be settled with the method of listing and drawing. Then, the strategy of using listing and drawing to settle the problem is the key point of this class. The lesson "I am a__________ child" aims at getting students to like to express themselves and love to write. At first, the teacher just makes his solo talking, but in the third class, the whole class is participants of composition correction. In the big change, the teacher also learns to change the focus from teaching task to students and from teaching plan to real situation.

Famous teacher and special-class teacher are the signboards and representatives of school. The two can be merged into one and also can be two separate ones with intersection. For school with continued culture, high popularity and reputation, the cultivation of famous teacher and special-class teacher shall be put on much stress. The following strategies can be used as reference: choose cultivating subject which is composed of municipal key members of subjects and leaders; adopt the system of double tutor that is to invite researcher to improve the researching ability of teacher and invite special-class teacher to improve the teaching ability of teacher; set up famous teacher's studio or research office to play a role in this procedure; set up relevant management system which shall be charged by specially appointed person; make good preparation that means school shall grasp opportunities to encourage cultivating participant to enter the list of special-class teacher, give up the management method of waiting negatively, developing undisciplined and cultivating blindly.

Schools in different developing stages may make a judgment of the level, category and supporting rate of cultivation project, and then make a choice according to the judgment. After that, school shall make a corresponding cultivation plan according to the choice. If it's not proper for conducting a cultivation of special-class teacher, it shall not be casted.

1.3.3.2.2 *Second is the strategy of general concern on students*

Strategy of general concern on students means that school must improve the attention on students efficiently. In China, it's very common that a school may have thousands of students. In cities, the least number of students in one school is 500. The larger the scale is, the more attention shall be paid. Every child deserves being cared, and they must get that. This strategy reminds school to overcome the following two phenomena:

The phenomenon of only caring about good students. In school education, there may be personality differences, intelligence differences and different family backgrounds which bring different developing opportunities to students. Naturally, different studying achievements cause different attention. Whether in class or in activities, students with special talent or perform outstandingly are more likely to be the examples and get attention. Once in a primary mathematic class, among the 42 students, one student got eighth time to answer questions but other 12 did not even get one chance. Besides, students who do not behave themselves may also get high attention. On the contrary, students who belong to neither good students nor naughty students are more likely to get less attention compared with other two kinds of students. So, the suggestions for school and teachers are: when planning the class culture index, ensure the conversation can be conducted completely; teachers shall be aware of "whom I haven't invited to answer the question"; pay attention to and change the state of speechless students using the approach of experimental research; cast more activities which are possible for all students to participate.

The phenomenon of strong girls and weak boys. Out of the physical factors, girls develop 2 to 3 years earlier than boys. On psychological side, it's the same. In primary school, this phenomenon is pretty common, which causes that girl is strong than boy. In primary schools, nearly all the school guiders are glib girls and boys seldom get the chance. It's necessary to do some researches and actions on how to allocate and balance teacher's attention on students from the angle of gender. Suggestions for teachers: allocate even opportunities of participating in activities to girls and boys; set up school-level project or apply for provincial or municipal project for researching the differences of gender and find out corresponding strategies; help boys to overcome obeying psychology and silent behavior; do not approve for girls' domineering behavior and so on.

1.3.3.3 Strategy of dealing with routine work

The strategy of dealing with routine work is the management strategy used by school to carry out the system culture construction. It includes organizational design, school system arrangement and ceremony management, etc.

The organizational design of the school is the framework that makes the school run normally; it is the first step in the construction of the system culture for dealing with routine work. Organizational structure includes two categories: the mechanical organizational structure and organic organizational structure. The mechanical organizational structure is also called the bureaucratic and administrative organizational structure adhering to the unified command. It has a chain of command with formal authority system and forms a width-limited, towering and impersonalizing structure. It has many regulations and includes functional organizational structure and distributed organizational structure. Any organization will not have the whole characters of the pure mechanical organizational structure in practice. Organic organizational structure is also called adaptive organizational structure or participative organizational structure, which has characters as looseness of structure, high flexibility, high adaptability, low complexity, low formalization, decentralization, etc. It includes linear and matrix school organizational structure.

Organizational design is the work and action procedure of those administrative staff for establishing or reforming of school organizational design and making structure decision. School organizational design refers to the work and action procedure of those administrators led by the headmaster to establish or reform the school organizational structure and make decision. School organizational design may refer to the following procedures:

1.3.3.3.1 *Establish organizational framework*

Lee G. Bolman and Terrence E. Deal put forward four kinds of framework models. They regarded the organization as factory (structural framework), family (human resources framework), jungle (political framework) and temple (symbolic framework) respectively. Most schools are willing to choose family framework for a school is a family. Human resources framework is based on psychology. They thought the organization was a big family which consists of a group of people who are needy, sentimental, minded, skilled but limited. They have learning ability and the ability to defend the old attitude and belief. The biggest challenge of this framework is: the school organization should be appropriate for people—to find the way that make every employee have work and feel good toward the work at the same time.

1.3.3.3.2 *Select the appropriate organizational structure*

Each form of the organizational structure has application condition, so we should select in accordance with the condition. Large schools are suited to choose the linear-functional structure and distributed structure. The linear-functional structure is a structure combining the linear system and functional system. It is the basic structural mode of the modern organizations. The linear system is set according to the principles of integrity and the functional system is set according to the specialization principle. Its advantages are concentration of rights, unity of command, well-defined duties and high efficiency; the defects are that the moving towards the linear system will cause the concentration of rights and biased towards the functional system will easily cause each department to act on its own. As a result, people involved do not know what to do. The distributed organizational structure is also called the divisional organization. This kind of organizational structure sets up various business departments under the leadership of the headquarters. Each department keeps separate accounts and has its own business management right. Its character is that each business department has full authority for management under the guidance of the general objective for the company, and it is the organizational structure mode combining the concentration of power of the fundamental policy with the decentralization of authority of the management. Its advantages are that it can inspire the enthusiasm and creativity of the business department and improve the adaptability and flexibility of the organization; its defects are overlapping organizations, wasting of resources, and each department is relatively independent and easy to become over independent. Simple structure is suitable for the smaller schools and new schools for the small scale makes the standardization unattractive, but it is convenient and effective by adopting the informal communication. The large schools who adopt the linear-functional structure can use the matrix structure at the same time. The matrix structure is a matrix composed of combined departments which are divided in accordance with projects or services and in accordance with functions. Its character is the duality combining the aspects of right, information, reporting relationship and system. Its advantages are task-centered, cooperated with each other and high flexibility; its defects are the complexity of the organizational relations and easy to have multiple leadership.[42]

1.3.3.3.3 *Design organic appended structure*

Administrators maintain the mechanical organizational structure of the school as a whole, and need to obtain the flexibility of the organic organizational structure at the same time. Effective selection is to append the organic organizational structure unit on the mechanical organizational structure. Task force and commission system are just the organic appended structure design.

[42]Cheng Xiaobing Educational Management [M]. Beijing: Beijing Normal University Press, 1999: 408.

Task force is a simple version of the temporary matrix to achieve a particular, definitely specified and complicated task, and it will be dismissed after the completion of the objective. Commission system is to deal with the selection of design of some problems across functional boundaries combined with background and experience of multiple people. Its forms are formal and informal and temporary and permanent. This kind of form can draw on the wisdom of the masses and make the decision tend to be reasonable; and inspire the enthusiasm of the organization members; and avoid grabbing all the power and promote the communication and coordination. But its defects are quite obvious: high cost, division of the responsibility, act arbitrarily of an individual or a minority of people, etc. The application of this form in the school is the common faculty and trade union congress and the committee of the aspects of the personnel, management, learning, foreign affairs, etc.

Ceremony management is very important. In order to make the school culture thriving and prosperous, we must make the important events ritualized. We need to create a full set of ceremonies and celebrations, which include conflict resolution ceremony, the welcome and training ceremony of new teachers, entrance ceremony of new students, work ceremony of teachers (giving lessons, preparing lessons, evaluation of classroom teaching, dressing, etc.), management ceremony (meeting frequency, layout of the meeting place, shape of the council board, seat sorting, etc.), award ceremony, promotion and retirement ceremony, etc. The ceremony makes the school culture expressed in a way full of cohesion. After schools reached a milestone achievement and work, they must hold ceremonies to make culture be fully demonstrated. Put values and heroic image deeply in the hearts and minds of teachers and students to form driving force for action. Without such expressive event, the school culture ranges from dispirited or flat to extinct.

1.3.3.4 Cultural material strategy

Cultural material strategy is a material culture strategy to manage school's material culture. It is the fourth emphasis and acting point of culture construction of school. It is the various physical facilities created by all the faculty members during educational practice and the concrete expression of spiritual culture of school. Applying cultural material strategy needs to focus on two matters.

1.3.3.4.1 *Design and materialize school logo*

If the school doesn't have a logo, then we can begin to design it. We could combine it with key activities and make the entire school to design or ask

Figure 1.7 Logo of the Chinese Academy of Agriculture Sciences Elementary School.

specialized corporation to design. No matter who the designer is, it will need to offer multiple alternatives. An appropriate solution must be selected and approved through full discussion, demonstration and recognition by the whole school and the solution must reflect and embody core values of the school. Once the school logo is confirmed, it can be applied to school office system and advertising system, such as PowerPoint template, cup, paper, pen, desk calendar, school bus, and school flag. Besides the logo design, there are many basic elements needing to be carefully selected, such as standard color of the school, standard font. Try to invite celebrities to write an inscription for handwriting of school name, school core value, school motto, school spirit, etc.

Chinese Academy of Agriculture Sciences Elementary School of Haidian District in Beijing (CAASES for short) derives its school motto—fresh life every day, based on this principle "growing up and schooling". Everyone in the CAAES regards everyday as a new beginning. There are new discoveries and new development every day. School badge looks like ear of wheat, full and elated. School badge color is inspired by the sun that gives growing energy to all things on earth, transparent and bright and clear. Color variations echo the whole process of growth for life from breeding to mature until sedimentary deposits. Each ear of wheat is a light, which aggregates into a bunch with cohesive force and tension. It symbolizes the process of construction of the school that students, teachers, students' parents and society all participate in. Campus culture is constructed in the ecological, natural and wise way under mutual close cooperation.[43]

[43]Liu Fang Growing up and Schooling: Our educational philosophy [J]. *Primary and Secondary School Management,* 2010(6): 53–55.

1.3.3.4.2 *Beautify the school environment*

It includes selecting color for school architecture and its implying meaning, naming for school buildings and roads, adding places of historic figures and cultural heritage, designing corridor, office and teacher culture, etc.

Childishness education of Primary School Attached to Capital Normal University has achieved the goal of "letting environment infiltrate childishness". Wall for inscription by celebrities and honorary wall is originally the best window to show the school. Later, the school replaced the content of these two walls from the point of "childishness education": inscription by celebrities in the east side of school gate has been replaced by warm greeting—"Hello, boys and girls", which is to show that school is open to welcome students to enter into the hall of knowledge every day; "Boys and girls, wish you a safe journey!" is the care ringing around children's ears and blessing sent to children who are going to leave school. Honorary wall in the west side of school gate has been replaced by "Smile wall of the whole school students". On the wall, a peace dove appears vividly flying high and smile photos of students are inlayed in the pattern of dove. Name of five buildings are respectively: "Tongxi Building", "Tongqu Building", "Tongyuan Building", "Tongyuan Building" and "Tongya Building". These names conform to the theory on school management and even embody profound humanistic care. Names of these five buildings become the most favorite "childishness works". We select three primaries and one connection color for color of these five buildings. Brown building represents fertile soil of school; green building shows live for life, standing for health, going up and innovation; orange building means harvest and delightful. The connection color is yellow, representing love and responsibility.[44]

Geographical location and campus of the Beijing Xicheng District NO.161 High School are unique; it is a corner of the Imperial Palace. Architectural style is simple and elegant and campus environment is elegant and pleasant. Based on the theory on school management that students' development is the basis, school strives to make each building, stone, water, garden, forest and road exert its using function, aesthetic function and educational function. This school initiates an extremely feature-rich "nationalized campus" construction project, plans education environment of the school and constructs characteristic landscape for the campus and highlights school cultural spirit of "a fire" and "a drop of water". The school carefully designed "the first

---

[44] Song Jidong, "Childishness Education" Promotes Happy Growth of Students [J]. *Primary and Secondary School Management*, 2010(2): 51–53.

landscape" made of combination of "pine, stone and bamboo" while entering to let teachers and students know essential feature of our humanness. There are three pairs of couplet in the office area. Each pair is full of philosophy and it is guide for the development of teachers and students. The first pair: Seek talent broadly, cultivate scholars widely; horizontal tablet writes "sincere". The meaning informs that we need to truly educate people with sincere labor to cultivate the whole students into men of tremendous promise. The second pair: More fickleness will lead to sorrow and regret, carrying on as before always harms all heroes; horizontal tablet writes "Constantly strive to become stronger". It lets everyone know that you should constantly strive to become stronger, steadfastly learn and work but also need think well and have the courage to bring forth new ideas. The third pair: The aspiration should be set far and should start from humanity, how to exert Tao and virtue lies in the love for people and matters and then you can grasp all practical skills of the society and can easily deal with various conversion of social roles; only after your intention is straight, you can cultivate your moral character, govern your family and rule the world. Horizontal tablet writes self-discipline and social commitment. Its meaning informs that learn to be a decent person before learning how to do things, morality, humanity and righteousness are more important than knowledge and only after your intention is straight, you can cultivate your moral character, govern your family and rule the world.[45]

School development is an involuntary and persistent development process by dint of promotion of internal strength and assistance of external force. Cultural consciousness is the highest state of school culture improvement and it is the inner dynamic force to increase school culture. The purpose of school culture improvement is to lead the school by culture, operate from a strategically advantageous position and make education behavior under domination of culture become the instinct of the school.

[45]The case is from speech of the headmaster, Ma, on the "Ma Jing headmaster's thought seminar of running a school" in Beijing Xicheng district of December, 2008.

2

School Culture Improvement Model

School culture improvement is the overall work of schools, arduous and repetitive. The best way to improve the quality and efficiency of cultural improvement is modelling. A model, namely a pattern, is an organism containing many "parts". Stemming from the reality, a model is a kind of theory instead of a simple method, scheme or plan; different from common theories, the theory is a concise form to express theories and easy to operate. In terms of the manifestation pattern, a model is a design thought and framework to optimize the configuration of all elements in the system. School culture improvement model, i.e. the design thought and framework of optimal configuration for all cultural elements in the school system is designated to pursue superior and united school culture and achieve the maximum development of people and schools. Models of different types can be constructed pursuant to different logics and objectives. In this chapter, three kinds of practical models stemming from the practice of school culture improvement and with the research nature are introduced: tripartite cooperation model, school culture-driven model and four-step model. Thereinto, tripartite cooperation model is the external improvement model of school culture; four-step model is the internal improvement model of school culture; school culture-driven model works as the intermediary agent to connect the other two models.

2.1 Tripartite Cooperation Model

2.1.1 Cooperative Practice and Research Review

2.1.1.1 Background of tripartite cooperation

The demand on high-quality K–12 education is increasingly expanding and growing. To accomplish the improvement of education quality, we certainly need to integrate the human resources, policy resources, material resources and information resources of universities, governments and administrative departments for education, K–12 schools and all other sectors of society. The demand of school management on transferring from experience management and scientific management to cultural management is increasingly growing. The plan of driving the overall development of schools by culture has been included to the leadership and management schedule. It becomes an intense demand as well as a responsibility of universities, governments and K–12 schools to explore and research the improvement and development model for domestic K–12 schools. The cooperation of universities and administrative departments for education with K–12 schools facilitates the transfer of development orientation of K–12 schools. During the transformation of K–12 education from quantity and scale to quality and efficiency, it's indispensable for schools to adjust their development orientation. Students' school achievement will become one of the assessment indexes but not the only one. Such cooperation can guide the tripartite co-operators to jointly learn to establish a social partnership. The research and practice can only be carried out and accomplished with the mutual collaboration of the three parties, including the learning of language system, tempering of communication mode, proper information management, construction and test of improved models, etc. The research and practice establish a platform by the means of projects and integrate preponderant human resources to jointly make contributions to the connotative development of K–12 schools.

2.1.1.2 Cooperation practice

It has been a long history for universities and K–12 schools to cooperate with each other. Especially in recent years, the cooperation between universities and K–12 schools become closer and more extensive; the description of cooperative relationship between universities and K–12 schools at different levels also become increasingly diversified. In the pertinent Chinese literatures, scholars usually distinguish the relationship of universities and K–12 schools as cooperation, collaboration, partnership and symbiotic relationship. In the English literatures, there are also a large number of words to describe the relationship between universities and K–12 schools, such as partnership, collaborations, consortiums, networks, clusters, interorganizational agreements, collectives and cooperatives. In fact, it's quite difficult to strictly distinguish the cooperative relationship of universities and K–12 schools. The research in this book chooses the "cooperation", or "collaboration" sometimes. The two words are similar in meaning. To establish the cooperative relationship, the following three conditions shall be met at least:[1] there exist differences between co-operators; the interest of both parties shall be satisfied; both parties shall be fully selfless so as to ensure the interest of members of both parties. The first condition is an objective existence for the cooperation of universities and K–12 schools, but both parties shall take efforts to meet the other two conditions. The establishment of cooperative relationship between universities and K–12 schools isn't merely aimed at resource and information sharing but also achieving the mutual satisfaction of both parties' interest and bringing forth corresponding reform. The cooperation in this article mainly refers to the activities that universities (mainly normal universities), governments (mainly administrative departments for education) and schools (mainly K–12 schools) achieve the improvement and development of schools' organizational culture on the basis of common will and in the principle of equality and mutual benefit.

Historically, various countries in the world are all, to a certain degree, paying attention to the cooperation of universities and K–12 schools. Early in the late 19th century, John Dewey started experimental schools, providing opportunities for normal university students to conduct educational practice in these schools as well as for teachers in K–12 schools to study courses of various types in universities. This contains the thoughts of cooperation between universities and K–12 schools. In the early 20th century, Charles W. Elliott, Principal of Harvard University organized a joint conference of universities and K–12 schools, discussing how to improve the educational and instructional methods. Conventioneers advocated that universities should participate more in the improvement of K–12 education which was the beginning of direct cooperation between American universities and K–12 schools. However, in form, the cooperation between universities and K–12 schools can be dated back to a long time ago, but as regards the background, objectives and nature, the universities-K–12 schools cooperation' drawing widespread attention after 1980s is a little different from the previous attempt.

[1]Sirotnik & Goodlad. School-University Partnerships in Action [M]. New York: Teachers College Press, 1988: 16.

Since 1980s, in the process of coping with educational reform, more and more universities and school districts have realized that it would be difficult to solve various problems in teachers' education and school reform if they still maintained the previous independent operation mode. If we want to achieve results, educational reform should be a systematic action with multi-party cooperation and participation instead of a linear and unilateral effort. The publishing of *A Nation at Risk: the Imperative for Educational Reform* gave rise to another educational reform campaign in America. In this campaign, many scholars proposed that universities and K–12 schools should establish a mutual complementary and beneficial cooperation thus to ensure the common development of universities and K–12 schools.

In 1986, Carnegie Forum on Education and Economy published *A Nation Prepared: Teacher for the 21st Century* which suggested setting up a "clinical school" for the cooperative relationship of universities and K–12 schools with the view of ensuring the quality of teachers. In 1980s, John I. Goodlad put forward the "symbiotic" partnership.[2] In 1990s, some scholars proposed the so called "U-S alliance" educational research and development mode, advocating that universities shall be combined with K–12 schools, learn advantages from each other and jointly promote the development of educational theory and practice. Such mode was soon applied in many countries. The British "K–12 based" teacher-training mode and American "professional development schools" are most typical examples among them. Afterwards, more and more universities in China developed a wide range of cooperation with common K–12 schools.

Since 2003, Taiwan Departments of Education has launched the "curriculum and instruction deep-ploughing plan" and its second sub-plan "professional partners in cooperation—cooperation plan of universities and K–12 schools" to assist teachers in acquiring the ability in practicing new curriculum. This plan deems that the professional partnership can be divided into horizontal cooperation and vertical cooperation. For example, for the horizontal type, the professional partnership is established among schools or teachers; for the vertical type, universities and K–12 schools join hands in planning the advanced studies and thus establish the partnership to enhance the connection of theory with practice.[3] Through the cooperation

[2]Goodlad. Educational Renewal: Better Teachers, Better Schools [M]. San Francisco: Joeeey-Bass Publishers, 1994: 113–114.

[3]E-paper of Taiwan Ministry of Education [J/OL]. (2004-04-27) [2011-11-11]. http://epaper.edu.tw

of universities which train teachers and K–12 schools and in the principle of "partnership, voluntariness and mutual benefit", this plan is aimed at establishing a long-term professional development relationship. Besides, a research plan related to curriculum and instructional innovation was also put forward, in which the departments of education are required to grant specific subsidies of no more than 300,000 yuan to each case in each year and the audit of the subsidies will be conducted every three years. The last audit will be referred to in the verification and appropriation of subsidies in the next year. This plan is mainly designated to assist teachers-training institutions to deepen their understanding in the practical operation of schools so as to improve the teacher training quality and further the combination of the theory and practice of the teachers-training major in universities.[4]

The Chinese University of Hong Kong has been providing professional support for schools in Hong Kong and has accumulated over ten years of practical experience. Since the foundation of the Quality Education Fund in 1998, the Chinese University of Hong Kong has received the fund appropriation to carry out the school improvement work which includes "Hong Kong Accelerated Schools Project" (1998–2001), "Quality School Project" (2001–2003), "Quality School Action Project" (2003–2004) and "Quality School Improvement Project" (2004–2009). These are all integrated, interactive and organic comprehensive improvement projects which cover the important fields of school work ranging from school management to instruction and learning. The "Quality School Project" is a comprehensive school improvement project which requires the Chief School Developers[5] to come to K–12 schools to join hands with frontline teachers in school improvement. The Chief School Developer in charge of the project will first assist participant schools in the review of school situation to make schools fully comprehend their own instruction and management situation and then assist schools in setting a common goal, establishing team culture and systematically develop schools' self-perfection mechanism through interactive school-based teacher-training workshops of different themes. In addition, the Chief School Developer will also introduce various new instruction and learning concepts via the workshops so as to assist schools in developing

[4]Yang Xuan. Final Thorough Research on Grade 1–9 Curriculum Reform [J]. Educational Trend Guide (Taiwan), 2004 (8): 1–6.

[5]Chief School Developers refer to the professionals specially providing professional support and help for K–12 schools in the project. Its members consist of college and university researchers, senior training specialists from administrative departments, senior headmasters and teachers of K–12 schools.

their own features. Furthermore, the project will provide opportunities for trans-school cooperation, visit and exchange of diversified forms thus to let teachers from different schools learn from and share with each other and therefore to establish a quality culture circle to jointly provide interesting and effective instruction for students. The project helps schools developing community resources. The project especially attaches importance to the introduction of parents' resources and helps parents understand the new concept of 21st century education and reflect their roles in their children's growing process. Schoolfellows and students of schools in the same district will serve as the leaders of activities thus to cultivate their leadership skills (see Figure 2.1).

2.1.1.2.1 *School improvement trend*

The ultimate objective of the project is that schools can become learning organizations and keep improving themselves to provide quality education for students. Through cooperation, not only can K–12 school teachers improve their professional ability and improve curriculum and instruction, but also university experts get an opportunity to give full play to their special skills and knowledge and gain insight into the educational reality so as to solve the educational problems at schools by integrating social human resource and other resources.[6]

Compared with Hong Kong and Taiwan, the mainland China has less experience in the large-scale cooperation between universitics and K 12 schools. At present, the cooperation of K–12 schools with governments and universities in mainland China is mainly government-led cooperation and initiative cooperation. Beijing and Shanghai are the two typical areas as regards government-led cooperation. In 2006, Beijing started the junior high school construction project, joining hands with Beijing Normal University and Capital Normal University and investing 10.6 billion yuan to 32 middle schools in Beijing. It achieved good results. In 2007, Beijing launched the standardized construction project for primary schools. All districts and counties took active action to seek for cooperation with universities. In 2007, East China Normal University signed an educational cooperation agreement with Shanghai Pudong New District, the content of which mainly included: research on comprehensive coordinated reform for education, research on

[6]Wang Zhenli. One Way for Cooperation between Universities and K–12 Schools to Improve Education: Starting with Hong Kong Experience [J]. State Education (Taiwan), 2002 (10): 76–80.

education resource allocation, cooperation in running schools, teacher training, holding educational forum, guidance on instruction and research, project design and guidance on scientific and technological education in Pudong

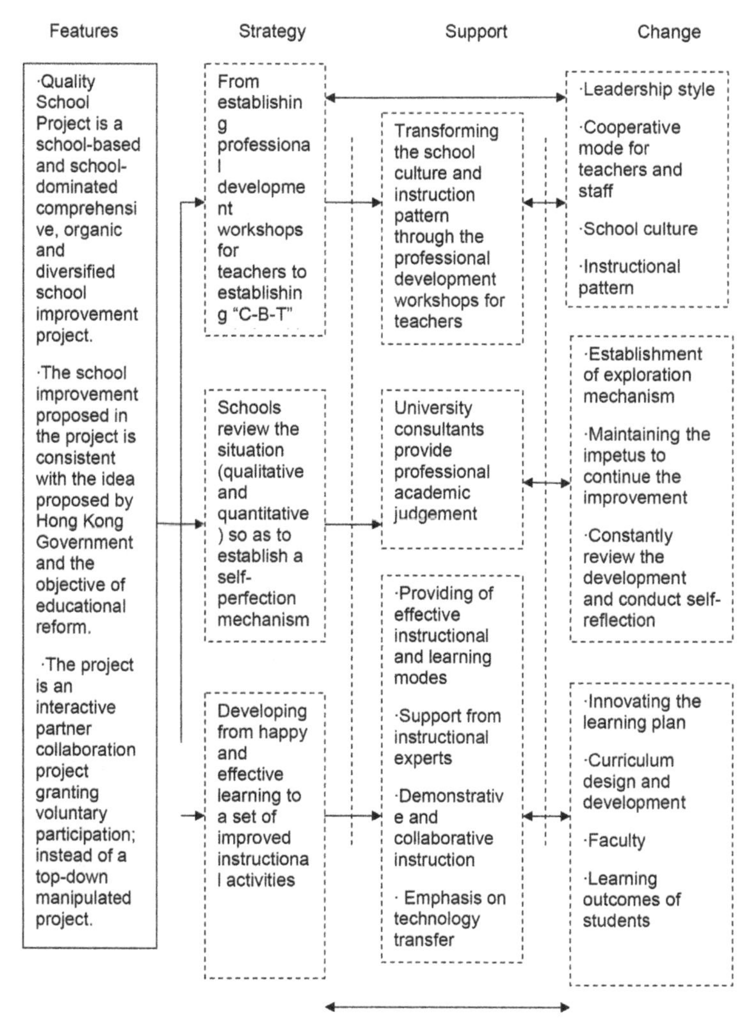

Figure 2.1 School improvement structure and strategies in Hong Kong "quality school project".[7]

[7]Zhao Zhicheng. Quality School Project and School Improvement [J/OL]. [2011-10-10]. http://translate.itsc.chuhk.edu.hk/ga/www.fed.cuhk.edu.hk/%7Eqsp/new/passage/w001.htm.

Youth & Children's Centre and information-based instruction research. The joint construction of Zhangjiang Experimental High School, Dongchang High School, Dongchang Middle School (South School) was one of the first kick-off projects. Professor Ye Lan of East China Normal University jointed hands with 55 key experimental schools (mainly in mid-east region) and over 50 non-core experimental schools in Shanghai to carry out research on new K–12 education theory and extension research projects. Regional popularization way was adopted in the implementation of the projects and "New K–12 Education" network as a platform to conduct experimental instruction and experience exchange. In the experimentation, the project steering group was the coordinators of the projects and the instructors of those experimental schools. University experts (including subject experts and basic theory experts) in the project steering group regularly went to different schools to conduct the main research by "attending a lecture, giving a lecture and commenting a lecture". No matter which school the experts went to, teachers from other schools in this district can come to attend the research activities. Through such cooperative mode, universities and K–12 schools carried out various activities concerning the reform practice.

Depending on the cooperative practice of universities and K–12 schools, five cooperation models can be summarized: teachers work at school while furthering study in the universities so as to increasingly deepen the comprehension theoretically and practically; in allusion to individual needs of schools, university professors research new instructional thoughts or strategies and seek for cooperation with schools interested in participating in the reform to jointly conduct study on the feasibility of the new thoughts and strategies; an university is associated with a group of schools to form a partner network to propel the overall educational reform and the network is characterized by reducing the hierarchical structure to make it easier to spread the news; university professors train reform agents to assist the reform; pre-service and in-service teachers do practice teaching. After the practice, the third cooperation mode was considered to be the most effective.[8]

Jon P. Wagner put forward there were three types of cooperation between universities and K–12 schools: data extraction protocol, clinical partnership and co-learning agreement. Li Zijian made comparison between the researcher-practitioner partnerships (see Table 2.1).

[8]Deng Weixian. Promoting the Curriculum Reform through the Establishment of Partnership between Universities and K–12 schools: Case Study on New Chinese Writing Project [D]. Hong Kong: Chinese University of Hong Kong, 2001.

Table 2.1 Partnership between researchers and practitioners[9]

	Data Extraction Protocol	Clinical Partnership	Co-Learning Agreement
Core issues	What is the nature of education and school education?	How do practitioners and researchers jointly improve the knowledge of school education and the practice at school?	What is the nature of education, school education and educational research?
Research process	Direct and systematic exploration; designed, hosted and reported by researchers.	Systematic exploration; jointly designed and reported by researchers and practitioners.	Reflective and systematic exploration; the continuous cooperation between researchers and practitioners stimulates the exploration.
Pattern and position	Researchers conduct reflection outside school; practitioners participate in activities at school.	Researchers conduct reflection outside school; practitioners participate in activities and conduct reflection at school.	Researchers and practitioners jointly participate in teaching process and school education system through activities and reflection.
Reform mode	Knowledge derivative from research stimulates policy and makes a contribution to improving teaching.	Researchers and practitioners do cooperative research, namely helping practitioners improve their efficiency.	Researchers and practitioners start complementary reform in their own institutes by extracting knowledge through cooperative research.
Experts' role	Researchers act as researchers; practitioners act as practitioners.	Researchers act as researchers and collaborators; practitioners act as practitioners and collaborators.	In their respective institutes, researchers act as researchers practitioners; practitioners act as practitioners-researchers.

In the background of school improvement, universities and K–12 schools tend to a more democratic and equal relation, and in the process of cooperation, the two parties adopt strategies of asking questions, consulting, discussing

[9]Li Zijian. Prospect of Curriculum and Instruction Reform: Dialogue about Enhancing Theory and Practice [J]. Peking University Education Review, 2004 (2): 75–87.

and developing. Researchers and practitioners are apt to clinical partnership Wagner advocates. Researchers help practitioners improve teaching ability and practitioners conduct self-reflection while participating in activities at school.

Colin Biott divided the cooperation between universities and K–12 schools into two modes: one is implementation partnership, namely expert mode which adopts strategies of giving, instruction, demonstration and implementation; the other is development partnership, namely mutual cooperation mode adopting strategies of asking questions, consulting, discussing and developing. In the former mode, universities send experts to teach and offer demonstration to teachers in K–12 schools, and teachers in K–12 schools change theories into a kind of teaching practice; scientific researchers in universities are regarded as authorities in terms of theory, and teachers in K–12 schools act as operating tools. This is a sort of self-approbation and preconceived mode. The latter one is a real equal cooperation mode in which the scientific researchers in universities, instead of requiring teachers to do strictly based on given plans, discover problems on the basis of the lectures given by K–12 school teachers, and then cooperate closely with school teachers and develop practical plans together through applying strategies of asking questions, consulting and discussing.[10]

After observing and studying the cooperation of 14 American universities which have existed for 3–5 years and K–12 schools, Carol Wilson and other people summarized five developing stages of cooperation. ① Organization. In this stage, establishers of partnership analyze the reason of cooperation, determine participants, draft principles of cooperation and management, and determine resources needed. ② Preliminary success. In the stage, participants hold meetings or seminars to discover their common interest and to know challenges they are facing. ③ Waiting for results. It is a lull after the preliminary success of the first stage, and participants try to achieve some real results through their working in such a period. People without enough patience quit and half-hearted people usually wait and see. ④ Main success and extension. After results of significance to participants come out, the basis of participation has been extended—including interested participants from various fields. At last, all participants will share their spare resources. ⑤ Mature partnership. Participants achieve their aims which they put forward. Community members and individuals from K–12 schools or universities take

[10]Ding Gang. Mutual Help: Internal Impetus for Educational Innovation [J]. Curriculum and Instruction (Taiwan), 2003 (2): 1–10.

a step on the way to evidently material reform. In brief, all aspects of real partnership are highlighted.[11]

Michael Fullan deemed the important conditions to maintain successful cooperation includes: ① context; ② rationale, which ensures appropriate goals of universities and K–12 schools, and the basis of successful cooperation; ③ commitment. Both sides of cooperation commit time, money and manpower; ④ structure, determining systems and procedures of communication, decision-making, disagreement-solving, etc.; ⑤ focus, that is to expand mutual sharing and to make plans specific enough to hold all participants together but maintaining independence and creativity as well; ⑥ process, that is to develop active interpersonal and professional relationship.

Based on the summery of his experience in participating in the cooperation for many years, Goodlad put forward ten pieces of experience: ① Duly handling the cultural collision that cooperation will inevitably encounter; ② reform of education institutes: as for cooperation of both parties, it is the universities (institute of education) instead of K–12 schools that are very stubborn and reject reform; ③ support and input from leaders: the leadership should have a clear understanding on their mission, vision etc. of their institutions, and provide supports for cooperation; ④ providing enough resources; ⑤ providing the real models of cooperation; ⑥ bearing the ambiguity in cooperation. As a human's activity, the cooperation cannot be conducted in accordance with the unchanged specific procedures. So it is necessary to bear the goal-free plan, behavior and evaluation; ⑦ avoiding the "speed-up" form. Though the cooperation will face pressure of submitting results, the structures and procedures of cooperation cannot be accomplished in an action; ⑧ correctly facing the debate on the process and results. The idea of dichotomy (such as theory/practice, quantitative/qualitative, saying/doing etc.), is inadvisable; ⑨ avoiding excessive structuring and insufficient structuring; ⑩turning leadership into empowerment and responsibility sharing.[12]

Weng Zhuhua believed that in our country, the universities, especially the normal universities have been establishing cooperative relationship with K–12 schools in many forms, such as "experimental school" or "affiliated

[11]Fullan, Erskine-Cullen, Ethne & Watson, Nancy. The Learning Consortium: A School-University Partnership Program: An Introduction [J]. School Effectiveness and School Improvement 1995, 6 (3): 187–191.

[12]Goodlad. School-University Partnerships and Partner Schools//Petrie. Professionalization, Partnership and Power: Building Professional Development Schools [M]. Albany: State University of N Y Press, 1995: 7–22.

school".[13] Since the 1980s, in order to adapt the trend of world education reform and the internal requirement of self-development, the domestic education sectors started encouraging the K–12 school teachers to work on scientific education research. Many normal universities or other education research institutes provided the professional support for the teachers of K–12 schools, developed partnership with K–12 schools and opened the long-closed road of communication for theory and practice of education. Compared with foreign countries, the present research of domestic academia on this thesis is relatively weak. The present research includes the following aspects: the first is the research on the cooperation between universities and K–12 schools which is designed to promote teachers' professional development. For example, Cao Taisheng and Lu Naigui respectively chose different partnership-based education reform projects in Hong Kong and Shanghai. By observing the daily communication and interaction between universities and K–12 schools, interviewing experts of universities, frontline teachers and headmaster and reviewing activity archives and reflection records, they probed into the teachers' growth and course of change in the reform situation.[14] The second is the research on cooperation between universities and K–12 schools under the background of the new curriculum reform. Teachers' participation in scientific education research is advocated in the new curriculum reform, but K–12 schools in our country often lack ability to do independent, normative, and scientific research on education. The new curriculum advocates school-based research. "Professional leading" becomes the hotspot in theory research. For example, Song Min started with the analysis on the international and domestic background of cooperation between universities and K–12 schools. She then expounded the necessity of cooperation between universities and K–12 schools under the background of the new curriculum reform as well as her understanding on the connotation of cooperative study.[15] Luo Dan started with motivation of cooperation, from the aspects of theory and practice analyzing three typical modes of cooperation between universities and K–12 schools under the background

[13]Weng Zhuhua. Analysis on Factors Influencing the Cooperation between Universities and K–12 Schools: A Case Study of Cooperation between Universities and K–12 Schools [D]. Shanghai: East China Normal University, 2003.

[14]Cao Taisheng, Lu Naigui. Partnership and Teachers Empowerment [M]. Beijing: Educational Science Publishing House, 2007.

[15]Song Min. The Background, Necessity and Connotation of Research for Cooperation between Universities and K–12 Schools [J]. Journal of Capital Normal University: Social Sciences Edition, Supplement of 2004: 202–204.

of the new curriculum reform: school-based instruction and research type, subject research type, type of teachers' developing of school, and conducting evaluation and reflection on the three modes separately.[16] The third is the study of obstacles and conflicts in cooperation between universities and K–12 schools. For example, Jin Zhongming and Lin Chuili believed that there were really many advantages on university-K–12 school cooperation mode, but there still existed some conflicts and contradiction, such as value conflicts, discourse conflicts, functional conflicts and standard conflicts. They thought it was conflicts that were pregnant with impetus to reform. Two parties participating in the reform in practice should mutually cooperating and carry out democratic discussion and effective integration so as to successfully promote school reform.[17] Niu Ruixue, on the basis of field study, through specific cooperation cases, analyzed the difficulties of cooperation in scientific research between universities and K–12 schools. These difficulties include indifferent teacher groups, loss of schools' support and outsider status of the researchers with no rights. In allusion to the difficulties of cooperative research, the author summarized the needs of cooperative research by combining with some experience of cooperation.[18] The fourth is the comparative study on cooperation universities and K–12 schools. It's common in the research on practice and experience of Britain and America and Hong Kong, China. For example, it has been over ten years for Hong Kong to explore the cooperation between universities and K–12 schools. In the relative researches, they described the practice and experience of Hong Kong from the view of the policy system, cooperation mode and mechanism, project implementation, and cooperation cases, which provides significant enlightenment for the cooperation between universities and K–12 schools in mainland China.

With the frequency of education reform activities, school improvement has become a key discussion and exploration topic in the academic world in recent years. Although many studies have realized that cooperation between

[16] Luo Dan. Motivation and Modes of Cooperation between Universities and K–12 Schools under the Background of Curriculum Reform [D]. Changchun: Northeast Normal University, 2006.

[17] Jin Zhongming, Lin Chuili. University: Potential Conflicts of Cooperation Reform between Universities and K–12 Schools [J]. Shanghai Scientific Research on Education, 2006 (6): 13–16.

[18] Niu Ruixue. Why Is the Action Research Stranded: Difficulties and Outlets of Cooperation between Universities and K–12 Schools [J]. Courses · Textbooks · Instructional Methods, 2006 (2): 69–75.

universities and middle and high schools is a means of promoting the school reform and the teacher development, while in China, especially in mainland China, the research of cooperation between universities and middle and high schools to facilitate the school reform is not enough in the theoretical circle. Based on this situation, the author has done a lot of researches and practical explorations. Since 2008, the author with her team has guided the cooperation and improvement of more than sixty schools, and put forward cultural improvement and three models for school improvement.

2.1.2 Tripartite Cooperation Model

The concept of school culture improvement can be distinguished between broad sense and narrow sense in the sense of tripartite cooperation. In a broad sense, school culture improvement refers to the activity and process of improvement conducted by universities, government and schools together. In a narrow sense, school culture improvement refers to the activity and process of improvement conducted only by key subjects, K–12 schools. Tripartite cooperation model is the external improvement model, a tripartite cooperation model. This model can be expressed in Figure 2.2.

In the practice of tripartite cooperation model, three parties perform their own duties. The government mainly refers to educational administration, which is the bond between universities and K–12 schools. Its job is to integrate educational resources to formulate policies on development and improvement of schools and supervise the implementation of these policies, to provide funds, to control its development orientation, to supervise the project progress, to evaluate the performance of universities and K–12 schools and

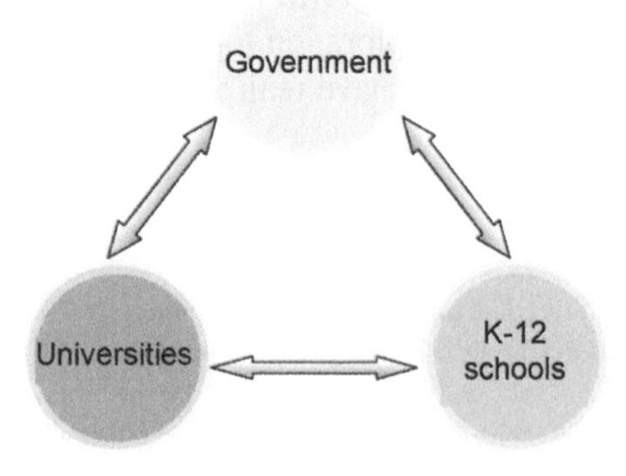

Figure 2.2 Tripartite cooperation model.

to be responsible for organization and contact between universities and K–12 schools.

Professionals in university are collaborators, consultants and researchers and practitioners. Their main duties and tasks are to formulate policies on cultural improvement of schools after consulting with the government and K–12 schools, to create the organizational mechanism and working system of tripartite cooperation, develop and improve all kinds of tools and techniques to advance the school culture, and to assist the K–12 schools to clarify their mechanism of school-running philosophy and practice to facilitate the development of schools. Universities contribute their advantages in human resources and information resources and thus expand the social capital of the project schools. Meanwhile, universities also benefit in data collection, research ability and professional development. In school culture-driven model, it's quite strict in selecting university members who should have two qualities and three capacities. The two qualities are responsibility and teamwork. Three capacities are the ability of communicate, the ability to put theory into practice and the ability to instruct. We can draw the conclusion that not all the staff in university is suitable for going into schools after tries and practices.

2.1.3 Cultural Tools for School Improvement

In his "Cultural-Historical Activity Theory" (hereinafter referred to as Activity Theory), Lev Vygotsky put that each conscious activity of humans occurred in specific situations. It is subject as soon as it has consciousness, which can be a single man or a group of men. No matter how many people they are, the subjects are initiative. From the viewpoint of the "Activity Theory", the school reform is a kind of goal-directed activity to advance the efficiency of school depending on the cultural tools, aiming to make schools be favourable for study. The cultural tools consist of language, symbol and other material instruments (such as chart, software, electronic information, etc.), which try to change the material situation so as to achieve the goals.[19] The initiators as subjects can be teachers and students from K–12 schools, or can be the collaborators. The initiative of new comers is limited to the rule, value and code of conduct but can be inspired by the material and spirit offered by schools.The improvement model of school culture created by us opens the

[19]Lu Naigui, He Biyu. Meaning of Initiators' Action: Exploration and Analysis on the Improvement Process of School Development Ability [R]. The Keynote Speech on the 3rd Cross-Strait Academic Seminar of School Improvement and Partnership, 2009: 6.

way to offer appropriate cultural tools to advance the school, which include the creation of organizational mechanism and the working system and the development of research tool, etc.

2.1.3.1 Combinational mechanism

Improvement model of school culture creates a set of modular organization mechanism and working system, including core workstation system, contacts system, workday system of expert group, lesson study system, system of mutual feedback on tripartite information, information sharing management system etc. we will illustrate the contacts system and tripartite information, information sharing management system in next section. Here we focus on the other mechanisms and systems that can be combined pursuant to the demands. The quintessence of tripartite cooperation is embedded in the combination and content of these mechanism and systems. The mechanism of teamwork requires that the management and improvement should be done by whole team instead of a single man or some people.

Core workstation system was created in 2008 when I hosted the first project meeting on normalization building of Beijing Fengtai First Primary School, which was applied to the 26 schools of third phase of the project cooperated with Board of Education of Fengtai. There are 100 primary schools in Fengtai where 26 principal workstations were completed based on the 26 schools with the view of making the project and the project schools play a leading role in the regional development. Every project school drives other 9 schools in different times. The strategy is allowing the project schools to run ahead of other schools and thus driving the development of other schools. The school being led must participate in the workday activity of project schools. They can inspect and learn from the activity, and also can be the main participators. Core workstations play an important role in driving other schools, spreading new ideas and enhancing the link between each other. The joint schools open to each other and go ahead hand in hand to create a forum of mutual learning and create an opportunity to communicate.

The workday system of expert group was also established in the project's preliminary stage in 2008. With the perfection of the model, the system becomes mature gradually. The workday system is divided into whole workday and half workday system. The whole workday system is mainly used in the preliminary diagnosis stage of the school culture development, consisting of classroom observation, headmaster report and brain storm, etc. (see details in next section). Half workday is applied in lesson study and experience summary stage. Teams continuously go to schools in the middle and later stage of the project pursuant to school demands. The team members include university experts, graduate students, instructors and researchers, headmasters of all project schools, leaders and managers of administrative departments for education. In 2012, the school culture construction project in cooperation with Baohe District of Hefei City of Anhui Province was launched, which was the first scale cooperation outside Beijing. In allusion to the new situation, the project team established the continuous working system of expert groups. That means expert groups have to continue working in project schools for days.

Lesson study is undoubtedly an excellent cultural tool. Lesson study adopts the modes that one teacher gives lectures on the same course or takes turns in giving lectures on different courses, or several teachers give lectures on the same course or takes turns in giving lectures on different courses, namely "there stages, twice reflections and behaviour tracking" mode.[20] The management process of lesson study includes plan formulation, lesson presentation (at least three rounds) and classroom observation, lesson discussion and communication as well as report writing, etc. The presentation way for the research results, instruction record, instruction research thesis, classroom observation and instructional diagnosis report and lesson study report, etc. Participators include university experts, teachers and researchers, graduate students, headmasters of project schools, teaching directors, and teachers, all teachers in the teaching and research groups and responsible personnel of administrative departments for education.

School culture improvement model requires that the information management must be complete. There are three types of information systems. The first one is comprehensive bulletin system for which graduate students are responsible. The project team formulated management rules. There is one theme for every semester. Its contents include new release in work days and its process and communication on class case studies. Second, the form of large keynote result report includes the assessment report collection of school culture development state, collection of plans for school culture development and collection of lesson study reports, etc. Third, the field notes of each project school shall be written by university experts and graduate assistants. All information is shared by universities, the government and

[20] Wang Jie, Zhou Wei, Gu Lingyuan. Reform on the Development of Teacher-Training Major [J]. Math Teaching in High School (Senior High School), 2006 (1): 110–113.

project schools. In accordance with the requirements of the administrative departments for education, the information also can be published and used in all schools of this region.

2.1.3.2 Tool kit

In the tripartite cooperation process, university experts also developed the tool kit for the assessment of cultural improvement state of schools, which mainly includes the following three modes.

The assessment indexes for the development state of school culture were set. The assessment standard for the development state of school culture consists of 4 Level I indexes and 14 Level II indexes (see Table 2.2).

The tools for classroom observation and lesson study were transformed. Lesson study and classroom observation cannot be separated. During the lesson study, the classroom observation technologies and tools we transformed and applied include: Information Table of Teachers and Classes, Instructional Schedule, Table of Observation on Students' Engagement in Study (see Table 2.3), Statistical Table of Question-Answer Behaviour Type and Frequency in Class (see Table 2.4), Observation Table for Instructional and Learning Strategies, Observation Table for Group Learning, Exercises and Homework Analysis Table (see Table 2.5), Verbal Flow Diagram, Teachers' Movement Map and Seat Map, etc.

Table 2.2 Assessment indexes for school culture

Level I Index	Level II Index	Assessment Result: Strong/Medium/Weak
Spiritual culture of schools	Core values	
	Educational objectives	
	School-running objectives	
	School motto and song	
	School logo	
Systematic culture of schools	Organizational structure	
	Improvement system	
	Improvement philosophy	
School behavioural culture	Leadership behaviour	
	Teacher behaviour	
	Student behaviour	
School material culture	School architecture	
	School landscape	
	Cultural facilities	

Table 2.3 Table of observation on students' engagement in study (one person) observatory

Observation Frequency (Once/5 mins)	Non-Engagement State		Engagement State	
	Number of Students	Percentage (%)	Number of Students	Percentage (%)
1				
2				
3				
4				
5				
6				
7				
8				
Total (number)				
Percentage (%)				

Table 2.4 Question-answer behaviour type and frequency in class (3 persons) statistician

Questioning Behaviour Type	Frequency	Percentage (%)
Conventional management questions		
Memory-based questions		
Reasoning questions		
Creative questions		
Critical questions		
Selection of question-answering way	Frequency	Percentage (%)
Roll call before raising questions		
Ask students to answer the question together		
Select those who hand up to answer the question		
Select those who don't hand up to answer the question		
Ask other students after raising the question		
Response to students' answer		
Interrupt students and answer the question personally		
Ignore students' answer or criticize negatively		
Repeat the question or students' answer		
Encourage and praise students' answer		
Encourage students to raise questions		
Students' answering behaviour type		
No answer		
Mechanically make "yes" or "no" judgment		
Reasoning answer		
Creative and evaluative answer		

Table 2.5 Exercises and homework analysis table (one person) statistician

	Exercise I	Exercise II	Exercise III	Exercise IV
Key points				
Cognitive level (memory, comprehension, application)				
Partial correctness rate				
All correct rate				
Main wrong answers				
Analysis on wrong answers				

Styles and forms of various results were formulated. To help school form a standard in the process of cultural improvement, the project team formulated written styles and forms of various results. These results include the writing forms of Assessment Report on the Development State of School Culture, scheme for school culture development, outline of report to headmaster, Report on Classroom Observation and Instructional Diagnosis, Lecture Giving Report, Less Study Report, etc. The elements and steps to write a complete class study report are: statement title, namely the theme of the study as well as a subtitle which should include the information of schooling stage, grade, discipline and lesson. In the analysis and discussion part, such questions shall be answered: how to determine the instructional objectives, how to teach by learning, generation and presupposition analysis, etc. In the study result part, it's required to expound the reached consensus and the problem worth further studying. The last part is the reference and acknowledgement.

2.2 School Culture-Driven Mode

2.2.1 Epistemological Basis of School Culture-Driven Model

The full name of school culture-driven model is the overall school development model in school culture-driven areas, known as school culture-driven model for short. It is an intermediary model which connects the internal improvement of school culture with its external improvement. There are three epistemological bases to construct such a model: cultural man hypothesis, man-cantered educational ideal and Blue Ocean Cultural Strategy.

2.2.1.1 Cultural man hypothesis

Cultural man hypothesis is the definition of humanity in the perspective of cultural improvement which puts that man is an animal influenced by organizational environment culture. The organizational environment culture is the independent variable while man is the dependent variable. The former determines what the latter will be, as the school culture determines what the faculty and students will be. In the independent variable, the core values of organizations are primary factors to influence man. The conceptual system cored with values has a decisive influence on man's behaviours. The development of organizational culture facilitates that of the man and vice versa.

2.2.1.2 Man-centred educational ideal

"Man-centred" philosophy is the origin of the second theory of school culture-driven model as well as the consistent and permanent nature and pursuit of education. "Man-centred" philosophy refers to the man is the centre of school, education and management. The nature of education is that teachers and students jointly create a bad or good life in which men grows up. Schools are quadrilateral sky carrying such a life on behalf of the state will. The man hereof includes not only students but also all staff of schools. Centre is a positional concept—the centre of schools, education as well as the society or even the whole China and world. Its profound value connotation is humanism: humans are the core, instead of margin, of education and activities and need to be highlighted; humans are the objectives instead of means of education, which is the true nature of education. That the man is in the centre of schools has the space significance, so the man is in the centre of education is of space-time significance. Early in 200 years ago, Rousseau had such a dream that the man became the subjects in the educational process. Humanism is aimed at exploring the students-based idea and achieving children-centred education. Teachers are the foremost at school because only when teacher are in the centre can students possibly be the centre of schools and education. After all, education is the process and behaviour given rise by teachers.

Who can be the centre of schools and education? People possess Chinese style and world insight. Such people dare to and can stand in the centre and will enter the society and step on the world core stages with Chinese flag and style in the future. To cultivate students who are able to stand in the centre, we must have such teachers who can and dare to stand in the centre. Schools that have such teachers and students must be cultural schools with profound deposits and clear values.

2.2.1.3 Blue ocean cultural strategy

"Blue Ocean Strategy" has been an eye-catching key word in the management circle of the world since 2005. The *Blue Ocean Strategy* co-written by W. Chan Kim and Renee Mauborgne has been translated into 27 languages since it was published by Harvard Business School Press in February 2005, which breaks the record of the press in outputting the international copyright. In the perspective of this theory, market can be divided into "Red Ocean Market" and "Blue Ocean Market" and the strategies related to it can be divided into "Red Ocean Strategy" and "Blue Ocean Strategy". The Red Ocean Market refers to the traditional industries and markets which are committed to defeat their opponents so as to maintain and expand their own living space. Due to the cruel competition, the market becomes bloodier and bloodier and its living space become smaller and smaller. On the contrary, the Blue Ocean Market is designated to create an emerging market without competition and its operation objective is reorganizing the value elements of buyers instead of defeating opponents, and thus creating a new market space "without competition" so as to completely get rid of competition and create a Blue Ocean. The analysis tools of Blue Ocean Market include: strategic layout and four-step framework. The former is a coordinate with the value elements of buyers as its abscissa axis and the commitment of different enterprises on these elements as its vertical axis. The latter include reduction, removal, increase and innovation. Blue Ocean adopts strategic actions as the analysis units and value innovation is thc foundation of Blue Ocean Strategy.

When applied to education, Blue Ocean Cultural Strategy is a blue originality, referring to the strategic conception and action of leapfrogging competition, creating a brand-new area and leading the advancement of regional education and schools. Schools often adopt diversified competition with a view of seeking for a long-lasting development and thus get stuck in the Red Ocean with fierce competition. To strive for a better future, regional education and schools are supposed to be adept in creating a Blue Ocean, namely creating a new growth space featured with cultural and value innovation. Besides, they are also required to get rid of the Red Ocean with exam-oriented education, leave their opponents behind, put forward new demands and stride to wider Blue Ocean. The Blue Ocean Cultural Strategy is the value guidance of the culture-driven model of schools, which takes in the core conception and strategic action of the vale innovation of the Blue Ocean Strategy with the purpose of facilitating the overall development and proposes the overall school development model in school culture-driven areas.

The implementation of the Blue Ocean Cultural Strategy should be based on tripartite cooperation model, depends on school culture construction projects and focuses on the culture-driven model of schools.

2.2.2 Orientation and Specific Description of School Culture-Driven Model

2.2.2.1 Orientation of mediation model

The school culture-driven model is orientated towards Blue Ocean Cultural Strategy, based on universities, schools and governments (USG Cooperation), taking school culture as a gripper to comprehensively promote a set of thoughts and operation frameworks for the overall improvement and development of schools, for the purpose of strengthening cultural solidarity and making schools develop towards a direction beneficial to teachers' professional development and students' healthy growth. School development refers to school culture promotion based on the original level. School overall development contains two aspects: a batch of schools in a region will have common development; the overall development of each school will be led by culture. The reason why the school culture-driven model is positioned as the mediation model of school culture improvement is that it functions as a bridge.

2.2.2.1.1 *Linking tripartite co-operators*

As a bridge, the school culture-driven model links university experts, administrative departments for education as well as K–12 schools together, and links the internal improvement model of school culture with the external improvement model, forming joint forces of school culture improvement. The school culture-driven model includes lots of innovative organization mechanisms and working systems, and the mediating effect linked thereof has got confirmation and development through the following two typical systems.

First, school contacts system: Each school sets up a core contact, who shall be a university expert appointed by a university team, with at least a doctoral candidate or postgraduate student as the assistant, as well as school principals and education administrators (including the director and deputy director of the education bureau, the section chiefs, contacts and researchers of K–12 education institutes, etc.), co-constituting a collaboration team for school culture improvement and construction.

Second, a system of mutual feedback on tripartite information, the school culture-driven model has established a system of mutual feedback on tripartite information including verbal feedback and report feedback. On the workday of an expert panel, the expert panel must give a 2~4 h verbal feedback to the schools. Within two months as from the working day, a written evaluation report for the development state of each project school shall be formed. The three parties shall sit together under the organization of the education commission, giving feedbacks to all school systems in the project about the report content. Each contact shall have a report for about 1 hour. Based on the evaluation report, the three parties shall jointly complete the planning scheme of school culture development, and then continue to convene feedback and exchange meetings.

2.2.2.1.2 *Communication study and practice*

The model itself is an operable framework between theory and practice. The school culture-driven model is a connection point communicating the internal improvement model of schools with the external improvement model, and it's also the mediation for school improvement researches and development practices, getting through the barrier between educational theories and practices, mainly for the following two reasons.

First, the school culture-driven model is derived from school improvement practices. In Dec. 2007, after the issuance of *Opinions on the Implementation of Standard Primary School Construction* of Beijing, all districts and counties put it into practice rapidly, actively formulating the development plan of standard primary school construction while seeking for cooperation with institutes of university education. In July 2008, Fengtai District Education Commission (Beijing) worked with the School Development Institute of Faculty of Education of Beijing Normal University to establish an Expert Guidance Group hosted by me, jointly carrying out the school effectiveness and quality improvement activity of primary schools in Fengtai District. The project team consists of 10 university experts, 10 postgraduates, 10 education researchers and 10 school principals. Fengtai Phase I Project only has 1 year. It has accumulated valuable experience and taken a key step to school improvement. The project team of Expert Guidance Group created a series of working mechanisms concerning contacts system, workday system of expert panel, core workstation system, lesson study system, etc. Particularly, the lesson study system has received great success, becoming the pinion gear and clear gripper of school culture improvement.

In Nov. 2008, the School Development Institute of Faculty of Education of Beijing Normal University worked with Haidian District Education Commission on the School Culture Construction Project of Standard Primary Schools, with duration of 3 years. Haidian Project was launched a half year later than Fengtai Phase I Project, continuing to and implement and improve workday system of expert panel, contacts system and shared information management system, with the addition of system of mutual feedback on tripartite information. The construction of School Culture Construction Project is: clearly explicating and operating school culture. School culture is an organization culture integrating schools' core values and distinctive features, including spiritual culture, systematic culture, behavioural culture and material culture. We break each culture into smaller variables (see Chapter 1). For easy understanding, synchronous with school work, we also use the operational concepts of schooling theory system (spiritual culture) and schooling practice system (systematic culture, behavioural culture and material culture) for realization and practice of school culture. So far, the bull gear of school improvement project has become clearer. The two clear grippers, school culture and lesson study, found an effective pivot for the school culture-driven model.

Second, the school culture-driven model is derived from academic researches and thoughts. During 2009 and 2010, by virtue of research subject application, I systematically organized several school improvement projects, clearly putting forward the school culture-driven model, applying for the subject "Experimental Study on School Overall Development Model of School Culture Driving Regions", with region as unit, systematically conducting the experimental study on the school culture-driven model. The entirety is relative, choosing a proper number of sample schools within a certain scope for experiment, using sample schools to drive other schools. Within a long time, it's allowed to conduct the experiment in batches, constantly modifying and improving the model according to the feedback information. Generally, a project needs 1–3 years. After the establishment of school culture-driven model, it has been applied and promoted in several projects, receiving good effect. These projects include: famous school brand construction project based on green education concept entrusted by the Faculty of Education of Beijing Normal University in 2011, schooling management culture construction project entrusted by Beijing Municipal Commission of Education in 2012, school culture creation project entrusted by Baohe District, Hefei, Anhui during 2012–2013, and school culture practice research project based on core school value system entrusted by Beijing Municipal Commission of Education in 2013. Adopt the school culture-driven model to wholly advance the experimental study of overall development model of regional schools. It can test the experimental effect of the model through practice, summarize regional education improvement model, provide reference model, experimental data and practice achievements for our country and each region to wholly promote school quality and effectiveness.

It shows that the construction of school culture-driven model is between school development and improved theory & practice. Based on the data collected in the model application and experiment, constantly test, modify and improve the model. It's not only an academic research model, but also an application study. The model construction is a process from theory to practice, practice to theory, and then being applied to practice.

2.2.2.2 Model description and variable analysis

School culture-driven model is the structure and framework for the optimal configuration of school culture elements and the promotion of school overall development, taking school culture and improvement as a gripper to fully drive overall school practice, making schools developing towards a direction beneficial to teachers' professional development and students' healthy growth, for the purpose of human and school development. The research elements of school culture-driven model include: the subject of driving, experimental variables and forms, the dynamic structure of driving model, etc.

2.2.2.2.1 *Subject of driving*

The subject of driving refers to the creation and application subject of school culture-driven model. Since it's a school culture-driven model under trilateral cooperation, universities, governments as well as K–12 schools necessarily constitute the subject of driving. In the aspect of school-running significance, K–12 schools are the independent subject and the key subject of school development while universities and governments are external subject and relatively secondary subject. Universities also serve as the role of mediation subject, responsible for putting forward, applying and modifying the model. The common goal of the three parties is to facilitate school culture improvement and development through school culture improvement.

2.2.2.2.2 *Experimental variables and forms*

The answer to the question relates to the research methods of projects and research subjects. This research chooses natural experiment method. Two questions must be solved for school culture research: first, how to define school culture; second, how to measure school culture. In Chapter 1, we solved the first question, breaking school culture into four operable variables. Here we'll solve the question of how to measure school culture and the driving model effectiveness thereof.

Natural experiment is one of the important methods for social sciences research. Evaluation research is a special type of natural experiment. It applies experimental logic to field investigation and research for observing and analyzing the stimulating effects in real life. Evaluation research is also known as project research and result evaluation aiming at evaluating the influence of social intervention, with influence of ideology. For the experiments of new teaching methods, new education modes, etc. can adopt the research methods of project evaluation. The approaches of information collection comprise the technical means like interventional observation, content analysis, questionnaire survey and interview. Social intervention refers to the actions adopted in certain social environment for obtaining some specific results. Evaluation research judges the result of a social intervention like a measure trying to solve a social problem. The educational intervention in this study is school culture-driven model, and the result is school and human development. This study adopts non-identical control groups in quasi-experimental design to test the model effect.

Natural experiment method relates to experimental variables. Experimental variables refer to independent and dependent variables for the experimental study and operation of school culture-driven model. The independent variable means school culture, i.e., taking school spiritual culture, systematic culture, behavioural culture and material culture as intervention means to influence project schools' cultural leadership and improvement and facilitate human and school development. The dependent variable means the development and change of human and school. The test criterion of the model effect is Development Theory Only, i.e., whether the culture driving model promotes school development depends on whether it promotes principals' professional development, teachers' professional development, students' healthy growth and school culture. The basis of change and development is the score change of displayed and expected values in the school culture evaluation tool.

Driving form refers to the mechanisms and means of school culture-driven model taking effect. In the sense of experimental study, it's the intervention means for independent variable. The first is content intervention. The intervention for school culture – independent variable can be converted into four operating variables: state evaluation on school culture development, program planning for school culture improvement, promotion of school culture improvement practice, as well as summary and refining of school culture construction achievements. The second is the intervention of organization mechanism and working system. What's the most effective method of combining and using excellent human resources and intellectual resources? Research and experiment require innovative and effective working mechanism. So to speak, so far, we are not lacking of people or money. What we are lacking of is the working form and mechanism for organizing people, money and materials, which is the system guarantee of school culture-driven model taking effect, including the aforesaid workday system of expert panel, core workstation system, contacts system, etc.

2.2.2.2.3 *Dynamic structure of driving model*

The school culture-driven model isn't anything emerged for no reason. The model establishment, the implementation and results of experiment as well as the process of feedback corrections are also the contents of research and practice, forming the model components. The model is dynamic and changeable. Though it's relatively stable, but it still needs improvement. The model can be tested. If necessary, it may able be re-established according to the research and experiment results. They are related to theories, able to be derived from theories. However, conceptually, they are different from theories. Based on the quantitative and qualitative data obtained in the study, the school culture-driven model may be amended. This part points to the feedback, reflection and amendment for the model. Besides, it also discusses and studies ethics and risk. The dynamic structure of school culture-driven model is as shown in Figure 2.3.

2.2.2.3 "One focus and multiple sub-domains" professional support strategy

The school improvement experience and model of Chinese University of Hong Kong (CUHK) is quite mature. Since 1998, it has received funds from the Education Bureau of Hong Kong more than once. CUHK has developed a series of complete school improvement projects in a form of university-school partnership. In the process of actually supporting schools, this model adopts the working mechanism of bull gear and pinion gear operation. The bull gear and pinion gear complement each other. The work of bull gear is to promote school effectiveness, to probe into education concepts and goals, as well teaching strategies, and to build joint group force and school culture. The pinion gear is to optimize teaching effectiveness.[21]

[21]Zhao Zhicheng. Reflection of University-School Partnership: Approach to School Improvement Plan [G]. Collection of Papers of the Third Cross-Strait School Improvement and Partnership Symposium. University of Macau, 2009: 29.

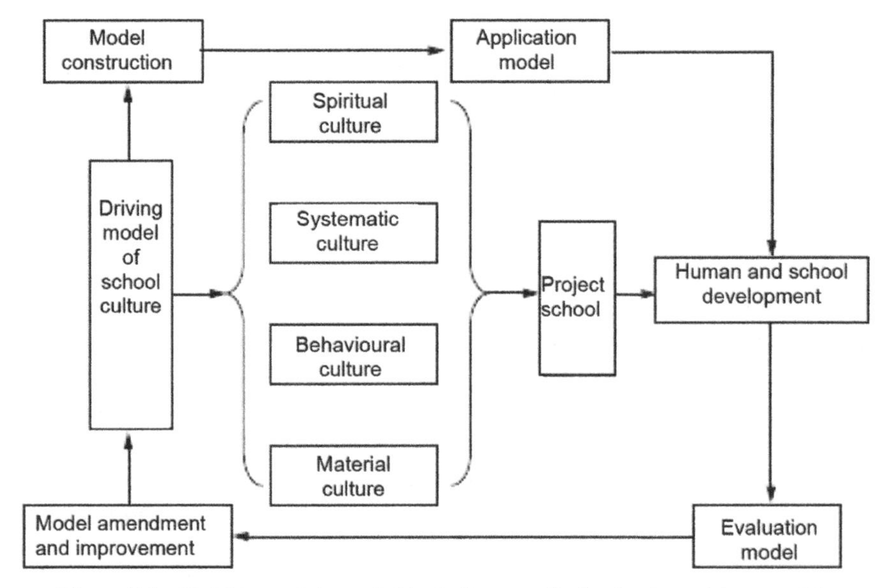

Figure 2.3 Variables and dynamic block diagram of school culture-driven model.

Whenever and wherever, the nature and function of education promoting human development and meeting positive demands of the society have never changed. School improvement plan and program are essentially the same. The school culture-driven model has constructed and practiced "one focus and multiple sub-domains" professional support strategy, which is a professional improvement point for giving play to the model, as well as one of the operating principles of the model. The implementation of the strategy adopts demand orientation and task driving. The demands and tasks of each school are different in different development stages. In the school culture-driven model, the demands and tasks of each school are of significance. Pure theoretical construction standing high above the masses is not allowed to act on schools. Instead, it's preferred to develop the culture concepts suitable for schools on the basis of culture facts, namely, following the thought of factual culture framework: fully respecting school development facts and traditions without imposing or adding anything; helping them get standardization, methodization, logicalization and materialization. We use partnership mode to assist school development instead of using pure data protocol mode.

"One focus" refers to a main gripper of school culture-driven mode, namely school culture and improvement. On this professional improvement point, based on school demands, university experts worked with schools to complete two tasks: state evaluation on school culture improvement and program planning for school culture improvement. The forms of two corresponding achievements are evaluation report on school culture improvement, and planning program for school culture development and improvement. "Focus" is the peculiarity of school culture-driven mode, and it's a bull gear and big gripper must be put into effect.

"Multiple sub-domains" refers to that school culture-driven mode has the practice improvement points of a plurality of sub-domains. These points are all within the operating variable scope decomposed from school culture, as significant areas of school work. The model defines "sub-domains" into several domains: management culture domain, curriculum culture domain, classroom and instructional culture domain, research culture domain, teacher culture domain, student culture domain, ritual ceremony domain, campus environment culture, etc. Within limited time, university experts can't pay attention to every aspect of these domains. The school culture-driven model adopts "one focus" + "one sub-domain" or "two sub-domains": based on the completion of "one focus", according to the independent demands of the project school, choose 1 or 2 aspects for professional assistance and improvement. For a 3-year project, the two tasks of "one focus" may be completed in the first year, and then the two sub-tasks may be completed in the second year. Generally, most schools may choose curriculum culture as well as classroom and instructional culture. Some schools may choose organization structure design etc. in management culture domain. In our project, most schools and districts chose to use lesson study to build a culture class so as to reflect and implement school culture. In the third year, evaluation, summary, refining, etc. for school culture improvement may be conducted. For a one-year project, the task may be quite arduous. In the first half year, it's required to accomplish the evaluation report on school culture development and the planning program for school culture improvement. In the last half year, it's required to accomplish the lesson study final report.

2.2.3 Application and Evaluation of School Culture-Driven Model

2.2.3.1 Application and improvement of the model

The development of school culture-driven model has undergone four stages, namely, exploration, proposal, improvement and application. Each stage has different understanding, thus the model development is different.

2.2.3.1.1 *Exploration stage*

2006–2009 is an exploration stage of school culture driving mode. Two critical events played roles. The first critical event is the 3 or 5-year "Junior High School Construction Project of Beijing" launched by Beijing Municipal Government and Beijing Municipal Commission of Education in 2005. It's a significant basic education improvement project aiming at developing a satisfying education for people, with an investment of RMB 10.6 billion. In the project, 32 schools in Beijing have been chosen to be the support focuses. Beijing Normal University and Capital Normal University have undertaken 16 schools respectively, responsible for providing "software" support. During April 2006 to June 2009, I was assigned to Hucheng Middle School, Chaoyang District. The process of participation in school improvement is a process of learning and thinking, starting to think about school overall improvement and project management problems in a regional unit: what's the cooperation mechanism among these universities participating in the improvement? What qualifies should a university expert have? What's the cooperation mechanism among university experts? And what mode is suitable for the inland school improvement of China?

With these questions, we got an opportunity to learn the school improvement experience of Hong Kong, and established the School Development Institute of Faculty of Education of Beijing Normal University in Jan. 2008, positioning education management on management practice of K–12 schools. In Dec. 2007, after the issuance of the *Opinions on the Implementation of Standard Primary School Construction* of Beijing, all districts and counties of Beijing actively formulated the development plan of standard primary school construction project. In July 2008, The School Development Institute of Faculty of Education of Beijing Normal University worked with Fengtai District Education Commission of Beijing, getting the project approval on the expert guidance group of standard primary school construction. This is the second critical event. In Fengtai Project Phase I, valuable experience has been accumulated. For short of experience, there're still some problems in project management: the project risk assessment is insufficient; the cooperation mechanism between the university experts and the education researchers is imperfect; when choosing schools, the district didn't put forward standards and requirements; the bull gear is the state evaluation on school development, which is too broad, without finding a suitable gripper. However, the lesson study as a pinion gear is very successful.

2.2.3.1.2 *Proposal stage*

The proposal of school culture-driven model benefits from the cooperation project of Haidian District Education Commission, Beijing. In Nov. 2008, The School Development Institute of Faculty of Education of Beijing Normal University worked with Haidian District, developing the school culture creation project of standard primary school construction. The project needs clear interpretation and operation of school culture. How to turn school culture into a specifically operable variable? How to turn academic discourse into practical discourse and achieve mutual communication thereof? How to take this as a gripper to facilitate the overall development of regional schools? So far, the bull gear of the school improvement project has become clearer gradually, combining with the little gripper – lesson study to implement class culture. And it works well. On this basis, the school culture-driven model was put forward.

2.2.3.1.3 *Improvement stage*

In July 2009, Fengtai Phase I Project was completed. Fengtai District Education Commission required continuing Phase II Project of Expert Guidance Group with us. And the duration is still one year. Fengtai District Education Commission chose 8 schools. Phase II Project made up the previous shortage and improved the model: the two large and little grippers have been unified in a sort of logic. School culture construction and lesson study became two different grippers and the cultural tools for school improvement.

The bull and pinion gears have operated in harmony. School culture is the ultimate concern of lesson study. The core school workstation has played the biggest role. During the workdays and lesson study, other 9 principals or directors of core schools earnestly and devotedly observed and participated in the project. They brought the lesson study methods back to their own schools and worked with education researchers closely and harmoniously. Phase II Project well divided the responsibilities of university experts and education researchers, arranging university personnel to take charge of bull gear, as well as lesson study organization and technical tool; arranging education researchers to undertake lesson guidance; arranging the university experts, education researchers and schools to complete the lesson study report. Besides, a return visit system was established.

2.2.3.1.4 *Application stage*

In March 2010, Shijingshan District Education Commission of Beijing worked with the Faculty of Education of Beijing Normal University to

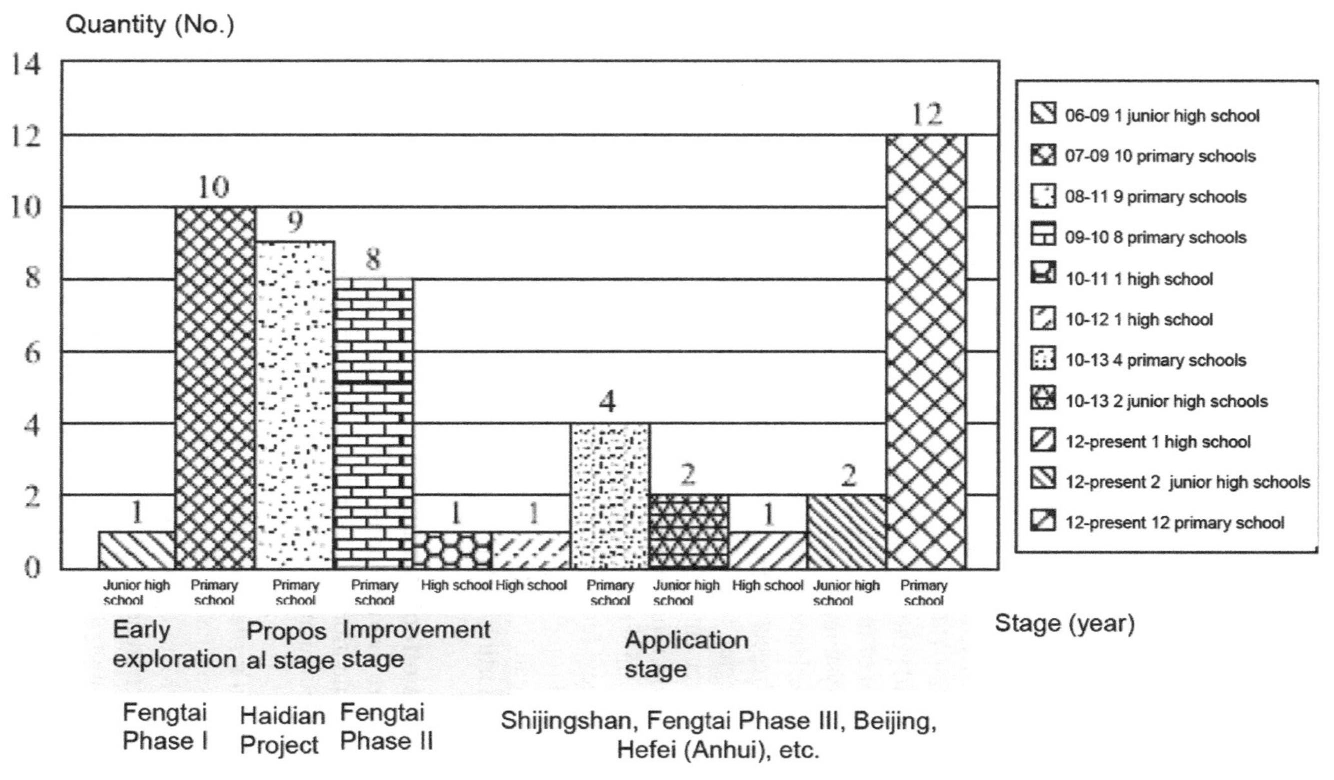

Figure 2.4 School culture-driven model development stage and school improvement quantity.

carry out green education development zone project. 5 major projects have been determined, in which, the "Promotion Project of Famous School Brand Construction Based on Green Education" chose 6 K–12 schools in the charge of the School of Education Management. Upon the requirements of Shijingshan District Education Commission and schools, the project team led by me decided to continue the experiment of school culture-driven model. It's a prepared battle. The significance of Shijingshan project is that it marks that the school culture-driven model entered experiment, application and promotion stage. This model was chosen for the following reasons: it highly matches and conforms to the connotations of green education concerning respect, sustainability, harmony and moderate development put forward by Shijingshan District Education Commission; the brand core of schools is school culture while the core of culture is the core value system of schools; the actual situation of most schools: disordered or excessively rich concepts, short of a logic system. At the same time, we've accumulated the culture improvement experience of over 40 K–12 schools. In Jan. 2011, the Expert Guidance Group Project Phase III of Standard Primary School Construction of Fengtai District began. In Jan. 2012, the school teaching management culture construction project entrusted by Beijing Municipal Commission of Education and the school culture creation project entrusted by Baohe District, Hefei, Anhui, kicked off, in which, 15 K–12 schools participated in school culture improvement. A new feature of these projects after Fengtai Phase III Project and Shijingshan Project is that the expert panel was organized to develop a quantitative research tool kit so as to conduct improvement and research simultaneously, starting with pre-test data collection.

2.2.3.2 Model's effect evaluation

The school culture driving mode evaluation can be divided into two parts: model's meta-evaluation and evaluation on model's actual effect. Over the several years, accumulation and degree of the two parts are different.

2.2.3.2.1 *Model's meta-evaluation*

The evaluation itself needs to be evaluated. The evaluation on evaluation is meta-evaluation. Meta-evaluation is a process of judging the effectiveness, feasibility, appropriateness and accuracy of the evaluation scheme or model. Good meta-evaluation can provide quality guarantee mechanism for evaluators. So far, the model's meta-evaluation is still blank. It's required us to keep trying.

2.2.3.2.2 *Evaluation on model's actual effect*

The actual effect evaluation of the school culture-driven model needs to start with three aspects, or use three methods of actual project management for effect evaluation. First, use quantitative data in evaluation. The evaluation subject is researchers. We've designed scales, questionnaires, classroom observation series tools, etc. of state evaluation on school culture improvement. During the several years, we've accumulated millions of characters of school culture improvement experience text, photos, videos, etc., as well as about 30 lesson study cases of 30 schools, with over 1 million characters of lesson study reports. And we've also collected a lot of data through observational tools and technologies. For school culture evaluation, we used scales and questionnaires to pretest over 30 project schools in the later stage. The posttest data haven't been accomplished wholly. We adopted the data of this part to determine school culture development degree and improvement effect. Second, use qualitative data in evaluation. The evaluation subject is also researchers. Use technologies and methods like content analysis to analyze and study the project school principals, teachers' discourse, manuscripts, etc., contrast the changes of discourse mode with structure in the comparison between the original draft of principal experience introduction and the planning draft of school culture construction, and weigh up the proportion and changes of experience discourse and academic research. Above all, the change of school running concept system can be used to prove the change of schools. Third, adopt the method of project evaluation. The evaluation subject is schools and administrative departments for education. Over the few years, the application of model has provided a systematic and profound angle for school improvement, bringing good effect, for example:

In June 2009, Fengtai Phase I Project School – Shiliuzhuang Primary School accepted supervision from Fengtai District Supervision Office. In the opinions replied, it mentioned: "The school has earnestly implemented the national moral education requirements, closely focusing on 'Wide Learning, Restraint of Etiquette, Repentance and Uprightness' to carry out school culture construction. Particularly, the activities such as etiquette education and traditional Chinese classics reading carried out by the school are popular among students, achieving remarkable effects." The school motto of 'Wide Learning, Restraint of Etiquette, Repentance and Uprightness" was determined under the joint discussion of the project team and the school.

In the opinions replied to Fengtai District Supervision Office, Fengtai Experimental Primary School wrote: Strengthen Campus culture construction, and improvement systems. Highlight the connotation of moral education, comprehensively strengthen formative education in multiple ways, carry out "Striving for Civilized Classes" activities, conduct "Civilized Class Creation Rating", and focus on Chinese classic poetry reading to implement sunshine education so as to facilitate healthy and happy life of students. Besides, systematic sunshine education is also one of the project results.

In May 2009, based on years of school-running experience, the project team put forward the concept of "Full Service and Precision Management" and the cultural system thereof for Fangguyuan Primary School. Over a semester later, the culture construction of Fangguyuan Primary School has made continuous progress. At the district S&T on-the-spot meeting on Oct. 23, 2009, the school-running supervision on Nov. 17, and the standard primary school construction stage meeting of Fengtai in late November, Principal Shen Ruizhi made a speech guided by culture concept, which obtained a good effect. The supervision expert Liu Zhanliang said, "I've never seen a principal understanding school matters so thoroughly before!" The acceptance appraisal of Fengtai District Education Commission says: "Full Service and Precision Management have infiltrated into each school work, with clear goals, key points, explicit responsibility division, effective measures, high operability and detectability, as well as high awareness rate of faculty." On Jan. 7, 2010, the director of Fengtai District Education Commission, supervision cadres and over 120 principals and leaders came to Fangguyuan Primary School (for the first supervision is very successful) again, held a "On-the-spot Meeting for Supervision Work" and implemented Scientific Outlook on Development. Focusing on "Full Service and Precision Management", Principal Shen Ruizhi made a report, bringing down the house. At the end of the meeting, participants were reluctant to leave.

On Jan. 21, 2010, the speech made by Zhang Fenghua, the Deputy Director of Haidian District Education Commission, at the Seminar of Nirvana Resort Convention Center of Beijing, is filled with rational thinking and perceptual spirit. She said:

"Participation in such an event is really exciting and full of passion. What I experienced is not a project but real education. It's a kind of arousal, which produces the essential meaning of education. I can experience the passion, wisdom and aspiration of project participants. The duration of each meeting was long. Everyone had an impetus to speak. The education commission is gratified at the project which has become more and more significant. Everyone is growing. We're learning from each other. The education atmosphere of culture projects is very thick. We've been seeking truth and goodness in the growth and development of schools. The conversation and sharing among educators are close to the essential attribute of education. The project has gradually proceeded from the exterior to the interior, becoming more and more zestful. It also shows that the process of education is a process of practice. At the beginning, we didn't have much expectation on the project. We know that the project development needs a certain process. However, now we see a surprise, feeling that school culture construction is close to the essence of education. Culture construction needs a framework to facilitate schools to carry out culture construction from spontaneously to consciously."

Guaranteed by the joint efforts of three parties and the effective mechanism, project schools obtained development to some extent. First, the school-running concepts, goals have become clearer, forming into the core value, which is the most distinctive in the school culture creation project. It's not only reflected in technology, but also reflected in thought. Sublimate experience into concept, get promoted with the aid of experts, and ensure that school culture construction develops from unconsciously to consciously. For example, the "man-centered" concept proposed by Zhongguancun Fourth Primary School, the Cuiwei education brand put forward by Cuiwei Primary School, the childishness education thought of Sijiqing Central Primary School (renamed the Primary School Attached to Capital Normal University later), the growth education concept of Primary School Attached to Chinese Academy of Agricultural Sciences, Peixing education of Peixing Primary School, etc., all contain the common wisdom of the project team. Second, systematic institutional framework of school culture development has been formed. For good school running, there's an essential stipulation and framework. Third, a project cooperation mode has been formed.

2.2.3.2.3 *Deputy director Zhang Fenghua considered*

"Schools are engaged in various projects. But our project is provided with a unique cooperation mode and project operation mode: goal leading – the team objective is to pursue for school culture development, guiding project development with cultural goals; task driving – specific stage tasks, each stage has a tangible task; setting items as required – each school has its own feature, and will find a focus according its own needs; in-depth discussion – not skimming the surface, not simple assertion, but comprehensive analysis, combining theories to guide practices has deep roots and lots of gains; follow-up strategies – after discussion, we should form a way and operation scheme, because without the process of implementation, discussion and brainstorming make little sense, and the key is the scheme implementation; continuous improvement – after follow-ups, seek for a promotion so as to see development and changes based on the original state; cooperation and sharing – we cooperate with each other and share the process and results. These achievements belong to us who have paid wisdom and emotions."

On Jan. 29, 2013, we held the Exchange Meeting of School Teaching Management Culture Project and the Effect Exchange Meeting of Fengtai Expert Guidance Group Project at Fengshan Hotspring Conference Center, Changping, Beijing. Liu Jianbin, Deputy Director of Fengtai Education Commission said, "The biggest achievement of Fengtai Expert Guidance Group Project is that it created a batch of wise principals able to systematically think of school development, and these principals can have outstanding performance even though they change schools." Director Zhang Fenghua of Basic Education Department 1 of Beijing Municipal Commission of Education appeared in the meeting. She was impressed by that. She said, "During standard primary school construction, since cooperation with our team in 2008, Fengtai District Education Commission has gone ahead in school culture construction, forming a batch of model schools of school culture construction, creating a batch of thoughtful principals able to run schools, promoting the rapid development of Fengtai primary school education."

2.3 Four-Step Model

Four-step model is a micro operation model of internal improvement of school culture, and an improvement process model as well. It shows how the culture improvement of a school is. The contents of Chapter 3 are based on the logic.

The four-step model of school culture improvement refers to the operational process and steps of school culture with a improvement unit as a cycle, including state evaluation on school culture development, program planning for school culture improvement, practice and promotion or program implementation of school culture improvement, and result evaluation of school culture improvement (see Figure 2.5).

2.3.1 State Evaluation on School Culture Development

2.3.1.1 Management process

After a school has become a project school, the primary task faced by the project team is to diagnose the culture development state of this school and the position in school culture spectrum so as to determine the strength of school culture. The three parties comprise a project team, namely evaluation team, dominated by the university expert group, working out the evaluation result of school culture development state based on the evaluation index system of school culture. This process is accomplished through three-stage investigation, multiple information collection, etc.

2.3.1.1.1 *Preliminary understanding of school needs*

In the project initiation stage, organized by the administrative department for education, university school culture creation project team and project school principals should jointly hold a kick-off meeting and a meet-and-greet event. Afterwards, the administrative department for education should communicate with the university team first to choose good project schools. The university team should coordinate with the administrative department for education, putting forward an improvement program of regional school culture, organizing a team, dividing the work. And it should take advices for the program from other two parties at the meeting, and announce the work division and contract list. Each school should prepare a 10-minitue written statement, introducing the school-running concept system, practice system and the construction process thereof, indicating the needs of independent guidance and the expectation to the project team, etc.

2.3.1.1.2 *Individual information collection*

The school culture-driven model adopts contact system. After the meeting, the fixed contracts sent by the university team to each project should take charge of the specific school work during the project. Each contract should

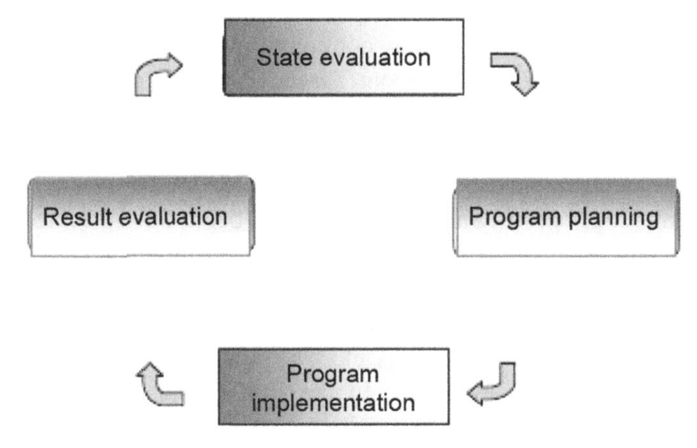

Figure 2.5 Four-step model of school culture improvement.

have a postgraduate assistant at least. Their task is to go to the school at least once or twice before the workday to get to know the school's overall development state, collect all collectable information by attending lectures, interviews, observation or questionnaire survey, and send the information to all members of the university team prior to the workday.

2.3.1.1.3 *Team workday*

The team workday is generally arranged 1–2 months after the project startup. According to the school work features, the three parties in the project team should negotiate with each other, generally taking April–May or October–November as the best period of workday. For the local projects of Beijing, generally a batch of schools will work together in the workday, a school per week. For non-local projects, generally 5 schools per week will work continuously and intensively. The participants of this workday include university experts, postgraduates, leadership team of school projects, work-station school principals, leaders of administrative department for education, education researchers, etc. The general number of people is between 30 and 50. The workday of state evaluation on school culture development is composed of 4 links, namely classroom observation, campus observation, interviews, principal reports and brainstorming. It generally lasts a day with large amounts of information. Classroom observation is to observe two lectures, which can be arranged in parallel with careful preparation. The subjects and teachers should both be arranged by schools. It's better to choose normal class. Campus observation is to observe the campus, teachers and students after two observing two lectures. The exercise between classes may be chosen. The interviews comprise group interviews concerning principals, secretaries, deputy principals, middle-level cadres, teachers, students, community representatives, parents, etc. Averagely, each interview lasts about 1 hour. The principal report needs 30–50 minutes. Interactive question and answer, teachers participation are both available. The last one is brainstorming. All university experts, principals of other schools and administrative departments for education present will air their opinions and suggestions. Since 2011, a link has been added prior to brainstorming, namely school culture diagnostic report by postgraduates. It's about 10 minutes.

All process information of the above three stages has complete records, including written records, photos, videos, etc. The whole management process is complete with sufficient evidence. In the link of state evaluation on school culture development, on the positive side, it's a discussion-based evaluation.

2.3.1.2 Discussion-based evaluation

Different from the supervision evaluation of the government, the state evaluation on school culture development is discussion-based evaluation, with its own features and requirements.

2.3.1.2.1 *Normal state evaluation*

The state evaluation on school culture development is a link of school culture creation project, as an evaluation in project management. The evaluation itself isn't the objective. As the primary work link, it's for improving and better managing school culture. Schools don't have any burden: with no need to prepare lots of materials, only needing to prepare a schedule and the existing publicity materials of schools.

2.3.1.2.2 *Low stake evaluation*

For the evaluators are university experts, they evaluate schools as a third party. Without stake relationship between the administrative subject and the school-running subject, schools have no administrative tense. 4–8 university experts all have doctoral degrees of management, pedagogy or psychology. It's a professional evaluation team, receiving a warm welcome from schools. The work belief of our team is: "One high and one low" – be low-profile in communication and behavior while be high-profile in professional level and thought. Each brainstorming is an on-site case training. The expert group will speak their own professional views from the perspective of cultural frames or cultural elements.

2.3.1.2.3 *Evaluation focusing on school culture*

The evaluation focuses on school culture, developing the evaluation index system of school culture, which is a high focus evaluation. Multiple methods such as interventional observation, multi-group interview and debriefing are adopted to prove the effectiveness and consistency of information. The evaluation on school culture isn't a thematic evaluation, but an overall evaluation from the perspective of school culture. The evaluation catches the most fundamental feature – school culture, which is beneficial to bring schools to culture improvement.

2.3.1.2.4 *Open evaluation*

Open evaluation means that the state evaluation on school culture development is sincere, comprehensive, encouraging and equal. Such openness is derived from three points. First, it's principle encouragement. The school culture creation project and plan abide by three principles: each school is unique; each school is respectable; each school has culture. Such principles let schools feel the sincerity of university personnel, trusting and accepting us soon to carry out the dialogue between universities and K–12 schools. The second one is the inclusion and openness of schools. On the workday, faced with the complete openness of universities, each school should no longer conceal any of its information, expressing itself honestly. Undoubtedly, with more openness, the school can have more gains. The third one is the equality of dialogue. In the brainstorming, each expert should fully air their views and opinions, and the principals of observation schools, leaders of the administrative department for education, principals of project schools, as well as everyone present could express their views and suggestions. All schools are open each other and have equal dialogues.

2.3.1.3 Result management

In the stage of state evaluation on school culture development, there're two result forms: the brief report of workday and the report of state evaluation on school culture development.

2.3.1.3.1 *Brief report of workday*

The brief report of workday is under the management of the postgraduate assistants of schools, and in the charge of the group leader, adopting timely collection and regular aggregation. Timely collection means that the postgraduates in the workday school should collect the required materials, including principal report manuscripts, interview records of each group and brainstorming records, accomplish workday news release within two weeks and submit the collection to the group leader. At the end of the semester, the group leader should collect all school materials and send them to all university team members and the postgraduates responsible for brief reports. In late December or late June before the exchange and summing-up meeting, the brief reports should be printed.

2.3.1.3.2 *Report of state evaluation on school culture development*

Based on verbal feedback of lots of information, the contacts of all project schools are required to negotiate with principals and schools repeatedly, absorb the suggestions beneficial for school development, get a clear understanding of the development state of school culture, find out cultural improvement problems, complete the first draft of school culture development report, submit it to all teachers for discussion, and finalize it after listening to their views and suggestions. It should be accomplished in late July or late December. The report has styles and formats for reference. It adopts school culture framework and often uses SWOT analysis tool. After this task, under the organization of the administrative department for education or university project team, the three parties will convene a special written feedback seminar. The university experts will report the evaluation on the schools that they contact with. Meanwhile, everyone will negotiate about the next work plan.

2.3.2 Program Planning for School Culture Improvement

After accomplishing the report of state evaluation on school culture improvement, this batch of schools will enter the planning program stage for school culture development. The contacts of university experts play a key role in this stage. In this stage, the following three-step improvement methods and flow can be referenced to answer what we should do.

2.3.2.1 Sorting – extracting

The information required to be mastered by schools and university contacts comprises: social mainstream ideology, education policies and intentions of the government, verbal feedback, written feedback opinions and suggestions of university team, existing problems in school culture development, existing historical traditions, cultural advantages and featured projects of schools, room for improvement of school culture improvement, situation of other schools, etc. On this basis, both parties negotiate with each other repeatedly with careful consideration, jointly teasing out or summarizing the preliminary thoughts for school culture improvement. They need to help schools do three things. First mobilize all available strength to ponder over spiritual culture (school-running concept system) clearly and extract it: considering that whether school core values, training objectives, school-running objectives, school mottoes and school logos are all clear and coordinated. Second, help schools match school-running concept system with practice system for integration, remedying separation and fragmentation. Third, accomplish the first draft of school culture improvement program. It shall be written by university contacts for accomplishment. For schools of mature culture, they may consider extracting core words to express school culture. Certainly, it's optional.

2.3.2.2 Consultation – joint discussion

After the completion of the first draft of school culture improvement and development planning program, it will enter consultation – joint discussion stage. The first draft shall be handed over to principals and schools, and schools will organize all or part of teachers of the subject group and classroom group for discussion, and organize students to discuss school culture part. They may also invite parents and community representatives to participate in the discussion. Make modification after fully listening to and collecting opinions and suggestions to form a discussion draft. To review the draft, they may start with the following questions. Can the program solve the loopholes and problems of school culture improvement? Can it clarify school culture? Have the feature points for carrying forward school culture been found? Can the plan turn into reality? If this program has been implemented, can schools have a better development?

2.3.2.3 Discussion – condensation

After discussion and opinion integration for long, modify the program to form a final draft accepted and supported by the entire school staff. It can be either written in the school development planning or passed by the Faculty Congress independently. In this stage, through certain precipitation, schools may have deep thinking. For further progress, a half workday of the expert panel may be followed once more to weigh and condense the program again. Even so, school culture improvement program can be accomplished neither overnight nor once for all. Instead, it can be a sign of interim understanding only. No matter how exquisite the program is, or how high the school culture level is, in the process of implementation, the program needs to be amended, adjusted and enriched continuously. In this sense, no one can be the authority. And no one has the final say. So to speak, school culture improvement is the eternal theme and focus of school development.

2.3.3 Program Implementation of School Culture Improvement

Compared with the planning program, program implementation and execution is the most difficult. Program implementation is a process for the further development of school culture practice. It's a long-term and hard work. A good program implementation requires schools to have strong executive ability. Program implementation may refer to the following management methods and flow to answer what we should do.

2.3.3.1 Clear grippers

Generally, the government and schools both wish the university experts would help schools have significant breakthrough in development, hoping to turn the program into reality rapidly. In a certain sense, this is a misunderstanding, because school is the key subject for self-development. However, the university experts may assist schools in further clarifying thinking so as to be clear about the two large and little grippers. The large gripper is the planning program of school culture improvement and development, led by school culture, thinking about school work comprehensively and systematically. The little gripper is a sub-item or sub-activity, as required by schools, focusing on one point within limited time. The strength may be chosen to carry forward culture and develop it into brand; the weakness may also be chosen to become a school feature with continuous enhancement. For instance, how to implement school core values with classroom culture? In the classroom in line with values, what's the code of conduct of teachers and students? In addition, systematic culture, student activity ceremonies or campus environment design may be chosen as a breakthrough.

2.3.3.2 Program simplification and marketing

School culture improvement program may be made into a full draft or simple draft. Describe the core of school culture improvement program in concise language within 5–10 minutes, simplifying the key parts and classic projects on a sheet of paper. An excellent program is a map, depicting the current state and future development direction of school culture improvement, as well as by whom and how the goal can be achieved. The simpler program may help achieve more thorough execution. A simplified program is easy to promote. Slogan shouting during morning exercises or class-break exercises may be adopted to spread school mottos or values; to be clear about the responsibility of each department and project leader helps monitor the implementation.

2.3.3.3 Specific responsibility to specific person

Program execution starts with target decomposition. Target decomposition is a process from responsibility target to work target, which can be conducted by hierarchical decomposition. Tree diagram and fishbone diagram are common decomposition and expression methods. All plans should be detailed and clear, with tasks assigned to each person. Progress, task nodes, completion stage, persons in charge, result expression, etc. shall be written clearly. After target decomposition, multiple task units will be formed. Work decomposition means that the major and deliverable results are decomposed into smaller task units for easy management, from top to bottom, from coarse to fine, until decomposed into the smallest controllable units, namely work package, and until decomposed into each person.[22] Post workbook and responsibility chart are common work decomposition methods.

Post workbook means to refine the duty provisions for fulfilling the task, as well as responsibility, operating process and operating methods, to quantize executive standard, and to refine and quantize inspection items. With documentation, the inspectors can minimize execution deviations and failures. Responsibility chart express the method of completing the individual responsibility of work unit in work decomposition structure in a form of table. In this way, it can lock responsibility, ensuring that school culture can be improved in reality while looking up at the starlit sky.

2.3.4 Result Evaluation on School Culture Improvement

Result evaluation is last step of school culture improvement flow, and it's an evaluation on the effectiveness of projects, programs and management. After the implementation of the school culture improvement program for a period of time, value judgment is required on the planning, implementation and result of the culture improvement program. Besides, it's also required to write down the practice and development report for school culture improvement program to answer how our work is going. The following two evaluation models may be referenced.

2.3.4.1 Process evaluation model

Process evaluation on school culture improvement refers to the evaluation on improvement process and procedures, including state evaluation on school culture development, program planning for school culture improvement, program implementation evaluation. The evaluation of each procedure has different standards and methods (see Chapter 4). The process evaluation model is as shown in Figure 2.6.

2.3.4.1.1 *Evaluation criteria and methods of state evaluation*

State evaluation has three evaluation criteria. First, sufficiency of background information collection, namely the investigation stage, whether the total information and culture improvement information of school development has

[22]Yu Shixiong. Winning in Implementation [M]. Beijing: China Social Sciences Publishing House, 2005: 54.

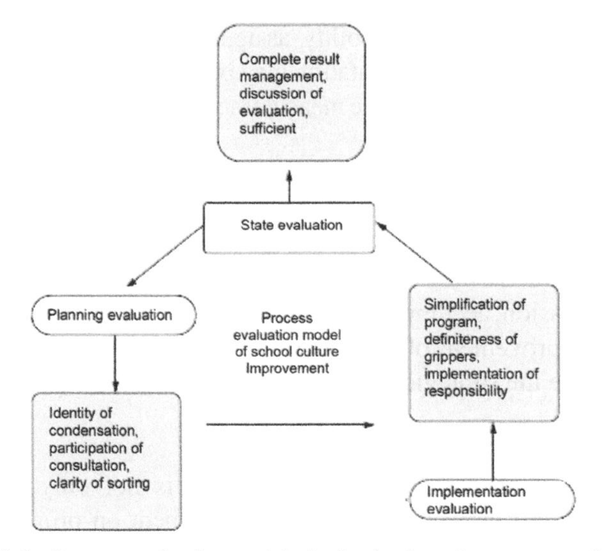

Figure 2.6 Process evaluation model of school culture improvement (1 Cycle).

been collected sufficiently, and whether the form is proper; second, whether the discussion-based evaluation is in place, whether it conforms to the principle of normal state, sincerity, openness and school culture focus; third, whether the result management is complete, and how the quality of reports and brief reports is.

The evaluation methods of school culture development are dominated by interventional observation method, questionnaire method and interview method, with various tools, including questionnaire, evaluation scale of school culture, interview outline, etc.

2.3.4.1.2 *Evaluation criteria and methods of program planning*

Program planning for school culture improvement has three evaluation criteria: first, clarity of sorting and extracting; second, participation degree of consultation and joint discussion; third, recognition and acceptance of school members for the condensed words. This evaluation may adopt meta-evaluation to evaluate the effectiveness, accuracy and feasibility of program.

2.3.4.1.3 *Evaluation criteria and methods of program implementation*

The program implementation evaluation on school culture improvement comprises three criteria: simplification of program and degree of publicity; whether the two grippers are clear; whether the work decomposition structure is reasonable with specific responsibility assigned to specific person. The evaluation data of program implementation may be obtained from the attitude or behavior change before and after the measurement program implementation of school culture evaluation scale.

The above is the process evaluation model of school culture improvement. It's a complete culture improvement process and cycle description. For a school, culture improvement is composed of many items, and it's also a management process consisting of a plurality of cycles which are linked. They are not only transmission, accumulation and succession of experience, but also amendment and improvement of management. The linkage of a plurality of items may constitute the strategic framework of school culture improvement.

2.3.4.2 Result evaluation model

Result evaluation on school culture improvement is not only the effect evaluation on culture improvement, but also a judgment on program effect and program implementation result. School culture improvement evaluates the effect of cultural improvement based on two objectives, namely promoting human and school development.

2.3.4.2.1 *School culture development*

Through one or several years of efforts under tripartite cooperation, do school culture and improvement have made positive progress? There are two judgment criteria: first, whether the school-running concept system of school culture is complete with logical consistency, and whether each core element is well-equipped with clear relation; second, whether the school-running practice system (including 9 elements in total such as systematic culture, behavioral culture and material culture) can become the carrier of core values; whether the relation between them is straightened out; whether a school culture tree can be drawn; whether the feature point of culture has been found. Fact listing method, questionnaire method and evaluation scale of school culture may be used to measure school culture development and changes. Meanwhile, the supervision conclusion of provinces, cities and districts are also the best proofs.

2.3.4.2.2 *Human development*

Human development comprises the development of principal leadership group, teachers and students. Attitude scale and lesson study method may be used to measure their attitude and behavior changes. Principal forum and teacher forum are both significant activities to inspect their growth.

3

School Culture Assessment

For leading the development and reform of school culture, it is necessary to learn about the state of the development of school culture, investigate and discovery its context. Seen as an indispensable and important activity that proceeds firstly in the construction of school culture, school culture assessment is crucial to the improvement and development of school culture. From this chapter, the book will meticulously analyze and respectively elaborate the model of the administrative process of school culture.

3.1 Characteristics and Subjects of School Culture Assessment

3.1.1 Analysis of School Culture Assessment

3.1.1.1 Definition of school culture assessment

School culture assessment is a territorial and specific one; therefore it follows the basic methods and procedures of assessment. Norman E. Gronlund, an American scholar, defines the assessment briefly as follows:

Assessment = Measurement (record and narrative of measurement) or non-measurement (record and narrative of quality) + value judgment.[1]

[1]Gronlund. Measurement and Evaluation in Teaching [M]. 2nd ed. New York: Macmillan Publishing Co. Inc. 1971: 6–10.

In other words, assessment is an activity of value judgment based on record and narrative. Record and narrative means factual judgment, which is an objective description of current situation, attribute and rules. On the basis of factual description, the value judgment is an activity of judging objective things integrated with objectivity and subjectivity based on the assessor's needs and wishes.

School culture assessment is usually called school culture evaluate or school culture audit. In English literature, "culture audit/assessment/evaluate" all are used. In Hong Kong and Taiwan, it is usually translated as "culture audit", but in mainland of China it is more known as "culture assessment" and "culture evaluation". There are also some scholars adopting the versions of Hong Kong and Taiwan, "culture audit". Although these three expressions are equally used in practice, they still have some delicate differences.

Compared with school culture assessment, school culture evaluation places more emphasis on the analysis and evaluation of the current situation of school culture, whose point is to discover problems and provide suggestions for further development. For school culture assessment, it not only requires judgment of the current situation, but also emphasizes the expectation for the future.

Compared with school culture assessment, school culture audit pays more attention to the examination and recognition of the fruits of cultural development with the purpose of evaluating the achievements of school culture construction. It usually acts as a summative assessment; relatively speaking, except for the evaluation of the fruits of school culture assessment, it could be also applied in other phases of school culture development with a meaning of formative assessment

Therefore, school culture assessment is not only a value judgment of the current situation of school culture development, but also a value expectation about the further development space of school culture; not only a judgment of the cultural fruits of school culture in a specific space, but also a dynamic assessment for the whole course of construction of school culture

Based on the comparisons above, the study chooses the concept "school culture assessment". It is an activity that comes out of the judgment of the development degree of school culture, a judgment of practical (already achieved) or potential (not achieved, but possible to be achieved) values of school culture construction, with an expectation of increasing the cultural values of the school.

As a specific form of assessment, school culture assessment has its own characteristics except for the general attributes of assessment. Firstly, school culture assessment is an assessment for the "culture" of school. Because of the complexity of culture, school culture assessment should have depth and breadth accordingly. The depth means that school culture assessment should not be limited to a direct and evident explanation. It is supposed to dig the deeper meanings reflected by phenomena. The breadth means that school culture assessment should not only have a span of time—penetration of the history, actuality and future, but also the range of space—inclusion of events, situations and people. Secondly, school culture assessment is an assessment for the culture of "school" It indicates that the school should have distinctive and outstanding features, and the content of the assessment should reflect the specific attributes of the specific object, the school.

3.1.1.2 Characteristics of school culture assessment

School culture assessment requires objective description and evaluation on the current situation, attributes and rules of school culture. However, as said before, due to the complexity of culture, it is hard to acquire the factual description of school culture and even harder to make a subjective judgment which accords with the objective reality. Because of this, it is very difficult to conduct school culture assessment. Although school culture assessment is hard to operate, some characteristics of culture still provide some guidance on the related methods and principles.[2]

First, culture has behavioral dependency. Culture is easy to be understood metaphysically, but it usually appears in events, situations and people. It is not a substantive. School culture may be embodied in the behaviors of teachers and students, in teaching, administration, teaching and research, even in a wall, a table, a tree, a photo, etc. Therefore, although culture usually contains a sense of vanity, it could be presented by the things attached to it. Assessor can acquire a perception of culture through observation of these external forms of culture. Culture has objective externalizations and carriers. This is an important reason why the school culture assessment can proceed.

To make school culture easy to assess, the assessor needs to systematically analyze different expressions of school culture, also its externalized carriers, and to ensure the entities could be included into the evaluation system, that is to say, to build a suitable index system for assessment. This will be particularly discussed later.

---

[2]Ji Ping. Self-evaluation of School Culture [M]. Beijing: Educational Science Publishing House, 2004: 38–45.

Second, culture has consistency. Primarily, culture is subjective reality, which doesn't vary from assessment or different conclusions of different assessors. Next, culture has publicity. It's a coexistence of values of people within the group. Through assessing different objects, the conclusions acquired could reflect the same culture itself. In reality, the consistency is the basis of school culture assessment to be credible.

In school culture assessment, triangulation can largely increase the credit degree of assessment. The basic principle of triangulation is to analyze and compare the observations and interpreting data on the related situations collected from various angels or positions. This method not only requires the assessor to use various techniques to study the same problem, but also spurs different people to analyze and evaluate the same phenomenon, problem or scheme from different angles. The consistency and differences of their opinions are both extremely important for the result of action research. Carl Ranson Rogers once proposed, the understanding of action should include subjective understanding, objective understanding and interpersonal understanding. The first one refers to the understanding of the assessor; the second one refers to the understanding about action of those who are thought qualified; the last one is to verify his own assumption by observing other assessors' conclusions. In school culture assessment, the selection of subject and object to a large extent influences the reliability and validity of the assessment. The subject and object of assessment will be elaborated later on.

Third, culture has symbolic characteristics. From words and actions, plants and trees, up to all kinds of ceremonies and rites, they are all carriers and representations of culture. The process of reading representations through carriers for the understanding of the symbols of culture is actually a process that culture is cognized and read. School culture assessment should be a process like that "a straw shows which way the wind blows". It needs to dig deeply each object and detail, to comprehend, analyze and judge the property of values of culture it presents, and to comprehend, analyze and judge the strength, moderation and measurement of culture it presents.

Last, culture has diffusibility. On the one hand, the diffusibility of culture means that the culture is not formed overnight, but requires a long term of accumulation, growth and combination; on the other hand, it means that perceiving and understanding culture can't be done at a time, which demands a frequent connection with it to perceive and deal with it profoundly. The diffusibility is supposed to be considered carefully in school culture assessment. A "fastfood" assessment would only make the school culture assessment flawed of indigestion, but a comprehensive one can make out the essence of school culture with the results of school culture assessment.

3.1.1.3 Significance of school culture assessment

As said above, school culture assessment is to judge the development state and the values of school culture. Its significance resides in judging the development state of school culture, finding and solving the problems. The motive of school culture assessment may come out from the understanding of the culture of school or the desire to improve it, and it also could be the necessity of the whole process of examination after the participation of external forces (such as culture-creating activity and culture-building project). The statistics are to be used to improve the school eventually. Aside from routine school culture assessment, most schools are doing school culture assessment because there are more or less problems in their culture and culture construction. "The organization in the text, in the actual operation, in the subjective understanding of administrators and in the subjective desire of managed objects, the degree of their consistency and self-control is a mark to decide whether the school culture needs to be evaluated."[3] An effective school culture assessment will greatly promote the construction and reform of school culture, meanwhile it also helps the sustainable and stable development of school.

The practice of school culture administration and construction cannot be apart from school culture assessment. The transformation and reform of school culture require analyzing the existing school culture and visualizing the direction of the school culture reform. It's necessary to be realistic, to find out and analyze the problems, for the healthy development of school culture. And the process and results of school culture construction also need feedback and improvement through evaluation. Therefore, school culture assessment has appraisal values for the development of school culture, which means to measure the level of school culture, and it also has discovery values, in other words, to find the highlights, features and existing problems through assessing and clearing the track of school culture reform. And it also has promoting values, that is to say, to promote the construction and reform by judging and analyzing, so it will boost the development of school culture eventually.

Carrying out school culture assessment benefits the healthy development of school. As a specific form of school assessment, culture shows in events, situations and people of school, school culture assessment somewhat is an

---

[3]Ji Ping. Self-evaluation of School Culture [M]. Beijing: Educational Science Publishing House, 2004: 28.

overall assessment on the school, which is in favor of discovering problems and promoting the development of school. In addition, it helps to clear up the cultural context of school and to understand its cultural system and strategy of construction. Because of this, the features of school culture will get developed and stand out.

3.1.2 Subject of School Culture Assessment

The subject of school culture assessment refers to persons who make value judgment on school culture. The "persons" here can be individuals such as assessment experts, principals, and so on, or groups or organizations such as assessment committees, education consultations etc., or the education administrative agencies like the Department of Education and Education Bureau. Based on the distinction between theory and practice, school culture can be described in two respects of system subject and executive subject.

3.1.2.1 System subject of school culture assessment

In this respect, governments, schools, intermediaries and customers all can be the subject, who may entrust, organize or participate in the assessment.

3.1.2.1.1 *Government*

In school culture assessment, the government as the subject refers to the education administrative agencies at all levels including the central and local education bureaus and education committees. Since it has administrative power and abundant resources, the assessment conclusion it makes is very authoritative and easy for people from all walks of life to accept. However, it may also exert disadvantages in terms of the assessment results and their use. The government needs to consider the interests of all aspects, so its assessment result brings different influences on different segments, and inappropriate handling will result in unjust consequences. The assessment conducted by the government belongs to external assessment. If there exist problems in the assessment staff, assessment plan and indicator quality, the assessment conclusion cannot be accepted by the school. In this case, the assessment result fails to maximize the function of feedback.

3.1.2.1.2 *School*

The school as the assessment subject has two advantages. First, it can conduct the assessment based on its full, objective and accurate understanding on the present situation and increase the efficiency in the assessment. Second, it is easier for the school to accept its own assessment result, leading to the immediate enforcement of relevant feedback. However, the assessment ability and level of a school is very limited, and people must be skeptical about it. At the same time, as the assessment subject of its own culture, the school is very likely to fail to compare itself with other schools. As a result, the school may be conceited about its culture, which is not good for the school culture development.

3.1.2.1.3 *Intermediary*

The intermediary as the subject meets the requirements of the assessment and is very professional. Therefore, in general, it can conduct school culture assessment quite well. However, we should consider whether it is qualified for such assessment. After all, unable to make a deep research, it can just analyze some relevant data. So it is difficult for the intermediary to understand the connotation of school culture.

3.1.2.1.4 *Customer*

The customer as the subject refers to students, parents, and schools of a higher level. They can share views from different perspectives, which, however, often deviate from the course as they are bias and temporary.

At present, the subject of school culture assessment in our country is primarily governments and schools. The education administrative agencies of some provinces and municipalities have already conducted school culture assessment and construction, which is always combined with projects of supervision assessment, characteristic school construction etc. At the aspect of schools, there are a few ones that spontaneously conduct the standardized school culture assessment, and the motivation for such assessment is weak. With the launch of the project of school culture construction, schools are enhancing their idea of culture construction. Once they were forced to go through culture assessment, but now they positively call for it. Taking the advantage of governmental project, they begin to combine the external drive and internal motivation for culture assessment.

3.1.2.2 Executive subject of school culture assessment

Obviously, such subject conducts the school culture assessment, including professional education consultation organizations, expert teams, and self-assessment teams with principals as leaders.

3.1.2.2.1 *Professional education consultation organizations*

They are a kind of the education intermediary, which can provide professional services for culture diagnosis. They have not only experts and professional investigators but also systematic assessment procedures and the professional templates of assessment report, which are helpful in completing the assessment fast and efficiently and finish the report. However, in China the majority of the organizations do not provide the service of school culture assessment, and the rest cannot conduct it perfectly. As mentioned above, they just analyze some superficial data. Besides, in their assessment, few staff from the school participates, making the assessment result less acceptable. Also, they will charge the school high fees, which adds financial burden to the school.

3.1.2.2.2 *Expert teams*

They are composed of experts in education, management, assessment and other fields, who have the professional knowledge, enough assessment experience, as well as the shortcomings of the professional education organizations. Their assessments are primarily entrusted and organized by governments.

3.1.2.2.3 *Self-assessment groups with principals as leaders*

In such groups, the representatives of the school are gathered under the leadership of the principal. They have obvious advantages in the spontaneity, deep research, participation degree, and acceptance degree and enforceability of the result. However, they may lack professional and systematic theories and standard manners and techniques. Involving its own interests, the result is not so objective.

In Table 3.1 below, the advantages and disadvantages of the three executive subjects above are listed as a comparison.

Table 3.1 Comparisons between the three executive subjects of school culture assessment

	Profession	Objectivity	Participation Degree	Research Depth	Acceptance Degree	Enforceability	Cost
Education consultation organizations	High	High	Low	Shallow	Low	Low	High
Expert teams	Medium	High	Low	Shallow	Medium	Medium	Medium
Self-assessment groups	Low	Low	High	Deep	High	High	Low

3.1.3 Types of School Culture Assessment

The object of school culture assessment is school culture. It can be understood in this way: "school assessment + culture" and "school culture + assessment". The former places school culture assessment into the assessment to form a comprehensive school assessment while the latter is a monomial assessment aiming at school culture. Therefore, school culture assessment can be divided into general assessment and specific assessment based on the status of school culture in the assessment. In a comprehensive school assessment, most assessment index systems are related to the assessment of school culture, such as the assessment of Blue Ribbon School Program in America and the performance assessment of schools in Hong Kong China. And this comprehensive school assessment is usually dominated by the government. So government is usually the system subject of school culture assessment and the executive subject is mostly a group of assessment experts consisted of educational supervisory personnel. "School culture + assessment" means the monomial assessment aiming at school culture. Assessments in this form can be done by different system and executive subjects.

To a large extent, it is influenced by the nature and type of the school culture assessment to decide the subject of school culture assessment. Regarding whether it is administrative, school culture assessment can be divided into the following two types.

3.1.3.1 Administrative supervision assessment

Administrative supervision assessment is an educationally administrative supervision conducted by the supervision department on behalf the government to make a value judgment on the administrative department of education, the educational management of the school, the quality and efficiency of the education and the development level in accordance with the national laws regulations policies and guidelines on education. Generally speaking, the system and executive subjects of administrative supervision assessment on school culture are usually the government and the assessment team of experts entrusted by the government respectively.

3.1.3.2 Non-administrative supervision assessment

Compared with administrative supervision assessment, non-administrative supervision assessment is a supervision assessment without administrative features. Moreover, it is often conducted by external and social groups, such as cultural institutions, parents, communities, foundations, with intermediaries or customers as the system subject and education consultants as the executive subject. At present, China has a few school culture assessments, thus the corresponding assessment subjects are not mature. In addition, non-administrative supervision assessment also includes the school's self-assessments which is an assessment activity independently conducted by the school to ascertain the development state and problems of school culture. Thus the system subject of this assessment is the school itself and the executive subject is usually the assessment group organized by the school. During the actual execution, it may be necessary to invite some assessment experts and educational consultants.

In actual school culture assessment, a frequently-used way is to combine administrative supervision assessment together with self-assessment of the school. In most cases, the school culture assessment dominated by the government is executed by a team of experts entrusted by the leading group of education departments at all levels For example, in 2008 the school culture construction program was launched in Fengtai and Haidian Districts, Beijing to regulate school culture construction in primary schools, with the school culture assessment conducted by the expert team from the School of Education Beijing Normal University. In recent years, more and more scholars advocate to launch self-assessment in schools with its own assessment team. They believe that the advantages of self-assessment cannot be replaced while its disadvantages can be mended through learning the theories and training on the assessment methods and procedures. Scholars like Ji Ping,[4] Zhang Wenqing[5] and He Zi[6] have introduced the methods and steps for the self-assessment of school culture. Yu Qingchen[7] also discusses the composition of self-assessment team of the school and he thinks that the composition of the team should be representative, that is, the team shall comprises representatives of different responsible groups; meanwhile, it shall also have an appropriate scale which means that the membership of the team should be restricted to 20. Also he points out that besides creating a new organization, the school can make use of the existing organizations or constitutions, for instance, the School Administration Committee, to conduct school culture assessment.

---

[4]Ji Ping: The Self-diagnosis of the School Culture [M]. Beijing: Educational Science Publishing House, 2004: 39.

[5]Zhang Wenqing: The Pointcut, Methods and Steps of the Self-diagnosis of the School Culture [J]. Elementary and Middle School Administration, 2004(7): 11–13.

[6]He Zi: The Self-diagnosis of the School Culture and the "Six Steps" for Transformation [J]. Elementary and Middle School Administration, 2004(7): 14–16.

[7]Yu Qingchen: Culturology of the School [M]. Beijing: Beijing Normal University Press, 2010: 158–159.

3.2 Content and Procedures of School Culture Assessment

Assessment is a kind of value judgment based on certain methods, skills and procedures. Since school culture can be assessed, then, what is the index system of this sort of assessment?

And how was it established? What kind of methods and procedures should be applied in the assessment? This section describes the content, methods, techniques and procedures of school culture assessment.

3.2.1 Content of School Culture Assessment

Content of school culture assessment is the index system of school culture assessment. It's composed of assessment indicators and standards, weight distribution, and assessment decisions.

Index system of school culture assessment is an assembly of various principles of assessment. Principles of assessment are essential rules for the content of assessment (that is, rules for the assessed attributes). It defines the scope of school culture assessment. Principles of assessment, according to their connotation and denotation, can be divided into several levels, namely dimension division. In respective dimensions, we can set up specific indicators and standards which are behavioral evaluation standards that have been disassembled, are measurable and operable. They are contents of assessment in accordance with the measurable or observable requirements, namely observation points.

3.2.1.1 Two design methods

There are two design methods for the index system of school culture assessment.

First, starting from the connotation of school culture, we can grasp the essential attributes of school culture and make the phenomenal appearance of the attributes as indicators. That means since school culture has many attributes, in order to make scientific assessment on it, we have to make in-depth analysis and grasp its essential attributes. For example, Mishra Denison put forward four attributes of culture in his cultural attributes model, namely adaptability, mission, consistency and involvement.[8] We take these four attributes as the first class indicators and set up three second class indicators to

---

[8]Denison. Toward A Theory of Organization Culture and Effectiveness [J]. Organization Science, 1995 (6): 204–223.

make assessment. It's the most concise and most effective measures to divide dimension in terms of essential attributes. But the difficulty usually lies in the analysis and extraction of essential attributes.

Second, starting from the denotation of school culture, we can take subculture or related culture as related indicators. For example, school culture SIS system assesses school culture based on the denotation of school culture and through the assessment of idea culture, behavior culture, visual culture and environmental culture etc. It is easy to establish an index system in this way, but it also has some defects. The main problem is a wide variety of indicators in this system.

Although it covers a wide range, it easily involves in some unimportant or irrelevant factors.

Both of the two measures above-mentioned can be seen in practice. By comparison, despite the diseconomy of the second method, it has been mostly used in primary and middle schools of our country because of its lower design difficulty. Almost all the index systems in the evaluation scheme of school culture issued by the administrative departments of education at all levels in our country have adopted the second method. While the majority of culture assessments in foreign countries, either corporate culture assessment or school culture assessment, have taken the method starting with culture connotation. Assessment systems designed by this means involve rich types and form various assessment models, which will be introduced specially in the later part.

3.2.1.2 Construction of weight set

If we say that index systems form the assembly of assessment factors, then, weight set represents the assembly of indicator relations. Weight of indicators reflects not only its status in the index system but also its relations with other indicators. Some techniques and skills are needed in the construction of weight set. Commonly used ones include Key-features Investigation, Delphi Technique and Analytic Hierarch Process etc.

Key-features Investigation only requires the assessed find the most critical or most distinctive factors in all the alternatives. It's an effective way to filter and entrust with the weight to indicators. Delphi Technique, also known as Expert Consultation, is a way to receive expert's "back to back" consultation in the form of questionnaire. After analyzing the answers, feedbacks are made before another turn of consultation. This should be repeated several times. Then people generally can agree on the results. Analytic Hierarchy Process compares each pair of indicators and gets the weight through matrix operations. It's a method to get the weight through mathematics.

3.2.1.3 Formulation of assessment standards

Assessment standard involves two respects. The first respect is the prescriptive nature of quality in the process of qualitative change. To put it simply, it means "the degree of requirements, excellence or complement". The second respect is measurement scale, which is our familiar "scale". From the perspective of the prescriptive nature of quality, it's hard to formulate absolute standards in school culture assessment, because standards are usually authorized by usage. People generally can agree on what kind of school culture is excellent and what sort of school culture is eligible. In practice, school culture assessment has usually adopted relative standards. One of the relative standards is to presuppose a healthy state of school culture. For example, Yu Qingchen[9] came up with the thought of democracy and equality, solidarity and collaboration, and forging ahead and sense of ownership, comparing the reality with the presupposed state of healthy culture and arriving at the result of assessment according to the gap. The following context will describe "Healthy-Pathological school atmosphere assessment model", which is a representative of this means. Another kind of relative standard firstly divides school culture into several attributes or respects and assesses the performance of school culture in each type or respect, so as to understand relatively outstanding (relative weak) attributes or respects. The below-mentioned "Assessment model of school culture types" belongs to this type. The feature of this type is its combination with school culture practice, making it easy to understand and operate.

From the perspective of measurement scale, the measurement of school culture assessment could be a yes-or-no categorical measure or a Likert-scale hierarchical measure.

3.2.2 The Methods and Techniques of School Culture Assessment

3.2.2.1 Two assessment methods

According to the basic principles of management science, human behaviors can be analyzed and the dependence of the culture reveals the explicitness of culture. Therefore, many western management theories believe that culture can be analyzed quantitatively. However, usually people would find out the truth

[9]Yu Qingchen: Culturology of the School [M]. Beijing: Beijing Normal University Press, 2010: 151.

that although culture can be externalized, it's still such a hard work to reveal its original face roundly and even harder to interpret its deep connotations, which has been already indicated by the symbolization and diffusibility of the culture.

Researchers of positivism methodology believe that culture is independent from its observers. Only if the reality is observed neutrally, the result of observation can be the basis of school culture assessment. On the other side, researchers of humanism methodology believe that culture is constructed, which is shown in forms of explanation. Culture thus needs to be comprehended and interpreted completely. From the very beginning of the culture assessment, there were two different assessment methods due to the two different methodologies.

One method is the qualitative research represented by Edgar H. Schein of the Massachusetts Institute of Technology. Schein and his fellows deconstruct the conception and deep structure of enterprise culture systematically by their long-time observation in enterprises. However, such method is so hard to evaluate objectively that it has received many criticisms. The other is the quantitative research represented by Robert Quinn of the college of business administration of University of Michigan. They believe that the organizational culture can be researched by setting observation point and from different dimensions. Hence, they propose some models for the measurement, assessment and diagnosis of organizational culture. But this method is classified into phenomenological method and believed that it can only make research on the surface of organizational culture rather than its deep meaning and structure.

The two methods are of the same importance in culture assessment. They are like two mountains towards and two rivers apart, of which either has its own merits and defects and neither can be isolated. Quantitative assessment is helpful to understand the current situation of school culture, saving both time and labor, which is like the "line drawing", with an excellent span and speed. And also it can provide data for some conclusions and make it easy to compare with other fields as well. However, the problem is that sometimes such method of school culture assessment could not reflect the real state of the school and with a low efficiency. And it's harder to find out the cultural meaning behind those explicit objects. Furthermore, difficulties for quantitative assessment are how to divide school culture into different parts according to the relevant theories (also dimensions) and how to decide the observation point and the observation target for each dimension. Different target system must result in a big difference in the conclusion of school culture assessment. The specific target system will be mentioned later in the book.

Qualitative assessment can satisfy the requirements of the assessment on depth, details and overall situation. It can dig out the cultural connotations and comprehend the organizational characteristics of the school more fully, which is more like "thick description". However, qualitative assessment is badly criticized on its credibility for it is greatly influenced by personal factors by stressing too much on assessor's explanation on culture. Moreover, qualitative assessment needs more time and energy and is with higher cost relatively.

In fact, in actual school culture assessment, methods used contain both the quantitative and qualitative research methods. That is, based on the quantitative research (like questionary inquiry and statistics of scales), the assessment would qualitatively analyze the information that is collected from field observations and depth interviews, so to get a relatively complete, objective and real evaluation result. One thing needs to be stressed is that the qualitative analysis method is undoubtedly of great importance for school culture assessment. It can not only verify and explain the values and meanings of the quantitative data, but most importantly can interpret the significance of a certain fact, detail or phenomenon to personal growth and school development. The value judgment generated from the research result is the argument of the assessment and assessment report while the quantitative data are just the ground of argument.

3.2.2.2 Several assessment methods and techniques

In order to do value judgment, we need to collect and analyze data for assessment by means of certain methods and techniques. There are many methods and techniques which can be used in assessment. We simply introduce several common methods and techniques for quantitative and qualitative research, such as questionnaire survey, interviewing method, observation method and text analysis

Questionnaire survey is to give out questionnaire or scale designed with the content to be investigated to the corresponding respondents and collect them back after filled, statistically analyze them and then draw a conclusion. The questionnaire survey has various advantages, such as involving a broad area, saving time and effort, convenient for statistic analysis. But it is bad for collecting deep and detailed data, and it is easy to be influenced by the level and mentality of respondents, and is difficult to control its reliability and validity.

It's the most crucial to compile the questionnaire and scale with high professional degree. School culture assessment needs to design tool kit for data collection. Since 2008, the project team for school culture construction

I led deeps into more than 60 middle and primary schools for school culture assessment and construction, developed series of assessment tool kits, including Assessment Scale on School Culture Development, Teacher Receptivity Questionnaire on School Culture, Questionnaire on Student Satisfaction Degree in School, Questionnaire on Parents' Satisfaction Degree, and also including interview outlines for schoolmasters, vice-schoolmasters, middle-level cadres, teachers, students, parents and community workers etc. Those tools are used in the beginning and closing phases for respectively comparing the pre-test and post-test data to ensure the progress of school and to evaluate the project effect.

After designing questionnaire and scale, we need to do trial testing to test the reliability and validity of the school culture assessment tools. As required, we need to repeat the process of design and test till the tools for school culture assessment are formally determined. The respondents of questionnaire and scale on school culture assessment should involve various groups, and schoolmasters, teachers and students shall all or partly (according to sampling method) fill in this questionnaire. Besides, relevant external public, such as parents, community and government, can also be the survey respondents. It is preferred that questionnaires are given out and collect back on site.

Data collected from scale and questionnaire are always quantitative and completed, so we often use various statistic tools, such as SPSS, EXCEL, to do some necessary processing.

Interviewing method is a kind of technology in which assessor gets data by means of interview. The interviewee is consistent with the extension of survey respondent. Compared with questionnaire survey, interviewing method is simple and highly flexible, can give full play to evaluator's experience and intuition, and can fully explore the meaningful content in this interview. However, interviewing method is time-consuming, and greatly influenced and confined by factors like the status of the respondent and the technique of the assessor. Besides, the collected data have certain subjective tendency.

Regarding whether there is an interview outline, interviewing method can be classified into three categories, namely open interview, semi-structured interview and structured interview. Interview outline usually comprises some overview questions, other than focusing on some specified behaviors. It is relatively more abstract and generalized, emphasizing on the whole cognition. For example, what is your opinion about features of school?

From the perspective of object quality, interviewing method can be classified into individual interview and collective interview. In school assessment, we can make specific arrangements in accordance with specific conditions.

During this interview, we can use interviewing log sheet (see Table 3.2) to collect assistant information, and the results are conducive to preliminary analysis. If we want to deepen the analysis, we thus need to try to translate the collected data (usually graphics, words and sounds) into words, and then analyze them by means of text analysis.

School culture assessment is more prone to use observation method. Advantage of this method is that the assessor can be on the site and understand the situation truly, but it also requires that the assessor should be a person with experienced professional knowledge and skills, otherwise, it is easy to distort the observation results when the observed adopt disguised behaviors to evade the observation.

When we are observing, we can use some relevant observation log sheets, as shown in Table 3.3, the log sheet on behavior observation.

It should be noted that either questionnaire data or materials of interview and observation should not be analyzed superficially; instead they shall be analyzed with cause analysis. We should deeply analyze causes and the cause of causes (as shown in Table 3.4) through behaviors and phenomena till we find out the implicit code and ideology behind those behaviors. Only in this way can we truly detect the essence of school culture. Therefore, the significance of qualitative analysis is evident.

Text analysis is not only a method that can be used to handle data acquired by interview and observation, but also be directly used in the qualitative research and analysis of school culture assessment. Through analyzing qualitative data, such as the regulations of school, annual reports, teaching plans, utterances and publications, we can get valuable materials by choosing, refining and arranging. Not only can the materials be used to test the results

Table 3.2 Interviewing log sheet for school culture assessment

Question	Status Quo	Reason	Feeling	Anticipation	How to Improve	Organizational Support
Interviewee						
Analysis						

Table 3.3 Log sheet on behavior observation

Time	Place	Behavior	Feedback	Preliminary Analysis

Table 3.4 Illustration for cause analysis[10]

Phenomenon or Question	Direct Cause	Indirect Cause	Background Cause	School Culture
Students have a poor ability in oral expression	Chinese teachers do not pay attention to training students' ability in oral expression	Teaching management: teaching and research group and lesson preparation group make not request for the cultivation of students' ability in oral expression Examination system: focusing on written test, not oral expression	Entrance system: further studies just rely on written examination score Assessment system: assess according to the written examination score	Implicit code of conduct: courses which are included in examination will be taught, vice verse. The teaching program focuses on examination. Utilitarian teaching and the tradition of ignorance of students' needs

acquired by investigation and research, but also reach a certain conclusion from the materials themselves. Text research has two aspects of tendencies. The more interesting we paid to superficial things, the more "deductive" we are. The more interesting we paid to profound things, the more "inductive" we are. Compared with other methods, text analysis can also save time and effort and expenditure, and not influence the staff related. The formation of its standpoint is influenced little by subjective factors, and it has a strong objectivity of materials, with a high reliability. The downside of text analysis is that the materials are seriously influenced by subjective factors of the original author, making it harder to distinguish their objectivity. In addition, materials integrity also influences the analysis procedures and results to a large extent.

In actual school culture assessment, we tend to comprehensively use various assessment methods and techniques. Taking "Primary School Standardization Construction" cooperated by the Education Commission of Fengtai District, Beijing Municipality and the School of Education, Beijing Normal University for example. In the earlier stage of the project, they used questionnaire survey to compile questionnaire and scale of school culture and

[10]Lin Yun. School Diagnostic and Development Basic Course [M]. Guilin: Guangxi Normal University Press, 2009: 113.

to investigate the headmaster, teachers, students, parents and community staff of the project school.

According to those data acquired, they sketched the cultural development state and cultural type of the project school. Then they launched an expert workday activity. They then used observation method to learn about the campus environment, the classroom teaching, and the activities of students and teachers etc. Again, they combined with interviewing method to interview the headmaster, middle-level cadres, and representatives of teachers, students, communities and parents etc. to collect more data on the school culture. Afterwards, they used textual processing to analyze the multivariate data collected in expert workdays, and used text analysis to diagnose the school culture and to complete the assessment report of school cultural development state. On this basis, they proposed suggestions for improvements of school culture and completed planning report of school cultural development.

3.2.3 Procedures of School Culture Assessment

For middle and primary schools, the "school culture assessment" here belongs to external assessment or one conducted by others in which college staffs the subject. The reasons for such manner are various. First, the improvement project entrusted by the Administrative Department of Education to colleges calls for it. Second, colleges are believed in terms of the assessment ability. Third, middle and primary schools are incompetent for self-assessment of its school culture. Besides, objective data involving the school culture development are necessary for colleges to improve schools of the project, and through this process, colleges are able to help middles and primary schools grasp the ideas, devices and techniques for self-assessment.

School culture assessment is a complex systematic project, but it complies with the basic steps of the education assessment. In the trilateral cooperation mood that is established and used in school culture construction, the procedures can be divided into three phases of preparation, implementation and application of results, with sixteen steps.

3.2.3.1 The preparation phase of school culture assessment

Before assessing school cultures, the preparation phase is very important. It consists of five steps and requires lots of preliminary works.

Step 1: determining the project schools of the project. After the Administrative Department of Education establishes the cooperative relation with a college, it should first choose the project schools, and determines the quantity and choosing standards. In this work, the opinions of the Administrative Department of Education prevail, and appropriate consultation with the team of the college is needed. The quantity of the chosen schools should be equal to that of the contact person of the college. Generally, there are two elements for choosing schools. First, the school principal is of enterprise and ready to advance the school. Second, the school can representatively show the education equilibrium of this region. After the determination of schools, information about these schools should be learnt, including analysis of social background, schools' needs and their expectation psychology. In analysis of background, ways of questionnaire and interviewing the key persons, and the tools of SWOT and PEST are often applied.

Step 2: determining the assessment team. In this step, colleges play an important role.

The host of the project is responsible for organizing the assessment team and a subsequent team, which are usually integrated. The host selects from a college 4–8 partners who are responsible and good at communicating and working with others. Each contact person deals with the specific works of a school, with a graduate assistant set up. In case the college team cooperates with the teaching and research department of its partner sometimes, it is suggested to organize a researcher team to conduct the teaching assessment. In the project by the author, college staff is responsible for organizing and guiding the lesson study, while researchers guarantee the lesson quality. Principals of schools learn from and communicate with each other, who are also the members of the team. After establishing the assessment team, knowledge and technology about school culture assessment should be studied and learnt together, so as to provide the most professional services.

Step 3: determining the work plan. In the mode of trilateral cooperation, school culture assessment is subject to school culture creation or the project of the expert team. Before the formal assessment, a working plan for the whole project should be completed. In the working plan, the project outline, work distribution, work mechanism, work manner, result forms and budget are carefully decided, so it is a guiding tool for the whole project. The one who accept the project should lead the team to finish this work. During this process, they need to communicate with the Administrative Department of Education and make some modifications.

The following is *the Work Plan of the Expert Team for Primary Schools Standardization Construction Phase II in Fengtai District*, which was completed by a team I led in the cooperation between the Education Committee of Fengtai District and the School of Education and Management of Beijing Normal University (BNU). It is presented here as a reference.

The Work Plan of the Expert Team for Primary Schools Standardization Construction Phase II in Fengtai District

By the work team from BNU for the Fengtai project
September 1, 2009

I. Overview of the Project

1. Project name: the Expert Team's Project of Primary Schools Standardization Construction Phase II in Fengtai District.
2. Time: launched in July, 2009, and finished in July, 2010, with a period of a year.
3. Construction purpose: this project focuses on the connotation construction and regards the acceleration of development as the core. It is expected to advance the efficiency in the direction, management, teaching, and scientific research of the primary school so that the balanced education development can be achieved in this district and the primary schools here can be cultural.
4. Project concept: every school is unique, and it has its own values and culture.
5. Work principles: operate in the form of project topic, construct in the form of topic guidance, and participate through the mode of problems and needs.

II. Project Tasks and Ultimate Results

Topic of the tasks: accelerate its development with school culture management.

Forms of the ultimate results:

1. The report of school culture assessment is in 8 copies (one copy for each school). Study the construction concept of school culture and its practice, analyze the advantages and disadvantages of the schools, make clear the educational philosophy to make it systematic and logical, and put forward the direction of further improvement.
2. The scheme of school development is in 8 copies (one copy for each school). Help the schools comb its systems of educational philosophy and practice.
3. Convene the forum for school development once or twice a year.

4. Carry out the activity of lesson study in the schools and assist the schools in finishing the report of the lesson study.

III. Work Mechanism of the Project

1. The expert team

In July, 2009, the expert team was set up, the members of which include educational experts generally from BNU (The number of these people depends on that of the experimental schools), teaching experts who are researches and special-class teachers at Chinese, Math, English and other subjects, and assistants (one for each school) who are responsible for the daily contact with the school, organization work, information exchange and supporting the technology.

In the daily work, each of the educational experts and teaching experts is responsible for contacting a school. He or she should bear the work of the teaching and education development, finish relevant diagnosis and assessment reports and planning schemes, and maintain the information communication with the school.

2. The combination of fixed appointed work and expert day

The way of fixed appointed work: each school has a fixed educational expert, a teaching expert and an assistant. They can unite as a team to conduct the regular work, together or in turn. In such team, the educational expert serves as the contact person, who is responsible for particulars, and the teaching expert should handle the teaching work.

The way of expert day: on the basis that experts independently conduct the work, the expert team holds the expert day at least twice at its school during the assessment period. All of its education experts, teaching experts and assistants participate in the expert day, diagnose and assess the school work in all respects, and provide advices and suggestions. They should ensure that they contribute all of their wisdom. The content of the expert day includes attending classrooms, communicating with principals, middle-level cadres, core teachers, students' parents and community representatives, listening to the report, and brainstorming the diagnosis and development of the school etc.

3. The combination of small team and big team

The way of small team: each school sets up a group of school standardization construction as a small team, consisting of from three to eight people. It must include a contact person (educational expert), a teaching expert and at least three people from the school. The principal serves as the leader of the team, who is responsible for particulars. The school designates a person to communicate with the assistant. He or she bears the work of collecting, disposing and spreading information including audio-visual materials and written materials etc.

The way of big team: all of the teams convene a meeting twice a term, at which experiences can be exchanged, problems can be solved, and the work thinking of next step can be formed. The meetings are planned by the big team combining the expert team and the said education committee.

4. The work thinking of the simultaneous running of big gear and small gear

At every school here, this work thinking is adopted in the school standardization construction. Big gear means the macro level of school culture construction, while the small gear refers to the micro level of educational and scientific research and teaching. The contact person is responsible for the big gear, and in the small gear, teaching expert handles the teaching work, the education expert deals with educational and scientific research, and the assistant assists the work.

5. The working platform

The form of Bulletin: the conclusion of the work circumstances of each school is edited by the assigned person of the team.

Forum for school development: it is organized by the Education Committee of Fengtai District and co-held by the School of Education and Management of BNU, aiming to exchange experiences, summarize problems and enhance the development ideas.

IV. Work Process, Action Plan and Staged Achievements of the Project

The project's work process, cooperated with the school calendar, can be divided into two stages.

1. From September 2009 to January 2010

The work focus of this stage is the assessment of school culture development state, including the assessment of the construction state of spiritual culture, system and management culture, behavior culture and material culture.

The job content covers deploying the project personnel and information, collecting and organizing all kinds of data and information, conducting classroom observation, and holds the Expert Day.

The staged achievements of work include school development assessment report (late Nov.), school development BBS (Jan.), and project bulletin (fixed period).

2. From February 2010 to July 2010

The system clarifies the practice system and philosophy of schooling and finishes the planning scheme of school culture development and the report on lesson study.

Step 4: determining the assessment program. After determining the work plan, the entrusted college project team is responsible for formulating a school culture assessment program, researching and developing assessment tools. The assessment program covers contents like the purpose, the index system, the method and process and the application of results of school cultural assessment. After making plans, the expert team put into a large number of efforts and money to develop assessment tools such as scale, questionnaire, interview outline and so on. In fact, this work and project are carried out simultaneously. It is not easy to form a complete tool suite in advance, as it needs mature and long-time research support and project support. After determination, the school development state assessment scale should be sent to the project school before site assessment. The data of pre-test shall be collected, so are the investigation reports on the school culture development state shall be formed.

Step 5: convening a conference. After preparing the above four "determine", the college team, the researchers and principals of the administrative department of education, the headmasters of the project schools and the graduate assistants shall jointly hold a multi-party conference. The project host is responsible to narrate the work plan and assessment plan. The participants are requested to give their own opinions. The headmasters introduce the cultural demands and basic information of their school and determine the assessment schedule on the spot. After collecting the preliminary information, the last adjustment and modification of work plan and assessment plan shall be put on schedule.

3.2.3.2 Implementation stage of school culture assessment

Assessment will be formally launched when the above 5 links are well prepared. The main task of implementation stage is to use the technology and method of school culture assessment to collect quantitative and qualitative data on site, and make a value judgment on the basis of the assorted and analyzed data. So the implementation assessment stage is also known as site assessment stage. The site assessment of school culture creation project is composed of 5 links: classroom observation, campus observation, headmaster's report, interviews of all the relevant personnel, and brain storming. They are also the 5 links of the team working day system of the project. And then add in the link of school culture development state investigation report to form 6 links.

Step 6: classroom observation. Classroom observation requires all the participants of the assessment activity to observe two lessons.

They can parallelly arrange several lessons and make well–preparations. Subjects and teachers are arranged by the school. The normal class is the best. The expert team can provide some simple classroom observation tools for the observers in advance. The school needs to provide the seating chart with students' names and the basic information of the teaching teacher of all the classes to be listened.

Step 7: campus observation. Campus observation is to visit campus, observe teachers, students and the setting-up exercises during the break, visit the janitor's room, library, reading room, canteen, toilet, corridor, professional classroom, laboratory wall, blackboard newspaper, and landscape etc. to obtain the true information of the school.

Step 8: interview. The interview objects include headmasters, secretaries, vice-principals, middle-rank cadres, teachers, students, community representatives and parents. The interview averages about 1 hour each group. At least one member of the college team in each group is responsible for organizing the interview. The graduate assistant is responsible for recording and tidying. Other headmasters and researchers can participate in any group at will.

Step 9: headmaster's report. The headmaster shall make a 30–50-minute report, wherein the participants including teachers can ask interactive questions and give answers. The report includes: self experience introduction, the stages of school culture development, the school's advantages and features, the school's difficulties and demands and so on. Before the headmaster's report, the college team shall give a report on the pre-test data collected before.

Step 10: report the investment of the development state of school culture. On every school working day, the culture development state investigation report of every school finished by collecting pre-test data is reported by graduate assistant, which costs 10 minutes. The report clearly presents the problems in culture development. The data obtained by observing and interviewing verify with these problems each other, which make up the origin of brain storming.

Step 11: brainstorming. Before calling it a day, there is another link of brain storming. All of the participants, such as college experts, headmasters of other schools, the leaders of the administrative department of education, instructors and researchers, can fully give out their opinions and advices. Every college expert shall make a 20–60-miniute speech in every place. They need to express their feelings and ideas of the day preliminarily tied on the site.

The Core Primary School of School Standardization Construction Phase III in Fengtai District

—the working schedule of Eastern Tieying No. 2 Primary School

Time: March 3, 2011 (Thursday).

Place: Eastern Tieying No. 2 Primary School, 259 Northern Tieying, Fengtai District, Beijing Municipality.

Working schedule

Time	Place	Activity Content
07:20	The classroom	The experts see every class doing morning reading
08: 00–09: 30	The classroom	Listening in class
09: 40–10: 05	The playground	Watching the exercises
10: 10	The meeting room	The headmaster's report
10: 40	The meeting room	The headmaster's forum
	The music classroom	The middle-level forum
	The science classroom	The teachers' forum
	The reading room	The community forum
	The arts classroom	The parents' forum
13:30	The meeting room	Brainstorming

The listening in class schedule

Time	Place	Subject	Teaching Contents	Teacher	The Introduction of the Teacher	The Listening Experts
	Grade 2, class (1), in floor 2	Chinese	*Animals' Fashion Show*	Jiang Lirong	38 years old, has teaching for 19 years, leader of the educational research group, one of Chinese backbone teachers in school, the third prize of "ShiHui Cup" contest in Fengtai District, senior in the elementary school	Zhang Dongjiao Xu Zhiyong
8: 00–8: 40	Grade 3, class (1), in floor 2	Chinese	*The Baby Iris*	Yang Fang	31 years old, has teaching for 12 years, leader of the educational research group, level 1 in the elementary school	Gao Yimin Ma Jiansheng
	Grade 3, class (4), in floor 3	Mathematics	*The Application of the Decimal Point's Moving which Changes Its Value, Eg. 7*	Zhang Yang	26 years old, has teaching for 2 years, leader of the educational research group, one of Math backbone teachers in school, level 1 in the elementary school	Zhouwei Zhao Shuxian
	Grade 5, class (2), in floor 4	Chinese	*The Tragedy of Righteousness Dog*	Zhao Aihua	38 years old, has teaching for 19 years, leader of the educational research group, senior in the elementary school	Yu Qingchen Yu Kai

(*Continued*)

Table: Continued

Time	Place	Subject	Teaching Contents	Teacher	About the Teacher	The Listening Experts
	Bungalow music classroom (1)	Music	*Papaya Is Just*	Chen Haishan	30 years old, has been teaching for 12 years, level 1 in the primary school, coming in September, 2010	Zhang Dongjiao Xu Zhiyong
	Grade 2, class (1), in floor 2	Art	*Colorful Butterfly*	Li Yuan	24 years old, has been teaching for 1 year, level 1 in the primary school	Zhou Wei Zhao Shuxian
8: 50–9: 30	Grade 5, class (2), in floor 4	English	*A Story Teaching—Unit 9 A football game*	Nie Jingwei	26 years old, has been teaching for 2 years, leader of the educational research group, one of the English backbone teachers in school, level 1 in the primary school	Yu Qingchen Yu Kai
	Grade 6, class (2), in floor 4	Mathematics	*Significance of Proportion*	Gao Yahui	32 years old, has been teaching for 13 years, leader of the educational research group, one of the Mathematics backbone teachers in school, senior in the primary school, coming in 2006	Gao Yimin Ma Jiansheng

The forum schedule

Place	Contents	The Interviewees	Experts
The Meeting room	The headmaster's forum	The headmaster	Zhou Wei
The music classroom	The middle-level forum	Director of moral education, director of teaching	Yu Kai
The science classroom	The teachers' forum	10 teachers	Zhang Dongjiao
The reading room	The community forum	5 representatives of the community	Ma Jiansheng, Zhao Shuxian
The arts classroom	The parents' forum	8 representatives of parents	Ma Yimin, Xu Zhiyong

3.2.3.3 The application stage of school culture assessment results

The application stage of school culture assessment results refers to the process of sorting data, getting results and finally applying them to improve the school culture. This process contains five links or steps.

Step 12: sort out the date to form a report. The data of preparation stage and data collected in field evaluation need a specific time for processing. In our projects, the requirements of all kinds of data sorting are the following: report the survey data in preparation stage on the working day; the graduate students sort out the interview recordings in about one week after the working day; the university contacts finish the report of school culture development condition assessment within two months after the working day; meanwhile, they shall finish the theme bulletin of school culture assessment.

Step 13: complete the report after classroom observation. Teaching assessment is a key point of school culture assessment. After the working day, the project team carries out teaching research activities in the project schools, using the tools of classroom observation to collect information. The university experts are responsible for the organization of teaching research activities and the work of projects management, and teaching researchers are responsible for the improvement of quality of the class. Under the experts' guidance, the school finishes the reports of classroom observation and teaching research.

Step 14: form a scheme based on the report. After finishing the report of school culture development condition assessment and teaching research, the university experts, by using the results of the report, finish cultural development planning schemes for every school, which will be applied in the improvement of school culture.

Step 15: feedback of the program appraisal. It contains two aspects. One is that the research data and results data have done by university project team all immediately return to project schools and the administrative department of education. Besides the written form of report of school culture assessment and cultural development planning scheme, the form of feedback can also be individual conversation, report-back meeting, and forum etc. The school can get the condition from the results of culture assessment, find the gap with the target state, and compare with other schools, so as to make the direction of school culture reform clear. The second is that after the assessment, university team needs to accept the assessment of the cooperated education administrative department and project schools. This assessment does not have to need to notify the university. The Administrative Department of Education calls on the headmaster of project school and teaching researchers to held a forum, discussing and assessing the performance of the university teams together, and returning the information to the university teams by different means so that they can continuously adjust plans and behaviors.

Step 16: self-improvement after the assessment. The evaluators or other expert teams need to evaluate the whole process and schemes of the assessment, they can follow the process of preparation, carrying out and feedback to reflect, or according to the schemes, reports etc. of the special things to analyze, evaluate its practicability, feasibility, rationality and accuracy.

3.3 Models and Tools of the School Culture Assessment

Based on the different value guidance and measurement indicators, the school culture assessment forms different kinds of models. All these models are nearly derived from or adapted from the organizational culture assessment field dominated by enterprise culture, so it will be necessary to introduce the models and tools of the organizational culture assessment.

In the last few decades, researchers have put forward many dimensions for the research of organizational culture. Stuart Alpert and David A. Whetten proposed to study organizational culture from the division between the whole and specific; Danny R. Arnold and Louis M. Capella put forward the strong to weak dimension and the internal to external dimension of concern; Deere and Kennedy put forward high-speed and low-speed dimension based on reaction speed and the high-risk to low-risk dimension based on risk tolerance; Breisach Ernst made the division between the participation to non-participation and the passiveness to initiative; George G. Gordon studied culture from 11 dimensions: direction, sphere of influence, organization conformity, the contact of the top management, individual initiative, conflict resolution, the clarity of the performance, the key point of the performance, action orientation, salary incentive and human resource development. Geert Hofstede focused on the power distance, uncertainty avoidance, egoism and heroism for the research of organizational culture; Kets de Vries and Danny Miller focused on the field about the disorder of cultural function, such as the paranoia, avoidance, bureaucratism etc.; Joanne Martin conducted the research about culture from its consistent and integration, alienation and conflict and cleavage and obscurity.

These researches lay the foundation for the models and tools of the organizational culture assessment, and some influential models and tools have got developed and certified. This book will introduce ten kinds of the most influential models and tools selected from numerous models and tools according to the following: (1) it has a clear theoretical basis; (2) it has passed the empirical test; and (3) it has great influence.

3.3.1 Models and Tools of the Organizational Culture Assessment[11]

3.3.1.1 Cameron and Quinn's enterprise culture assessment instrument (OCAI)

Cameron and Quinn believe the organizational culture got reflected through the ways of the values believed by the organization, dominant leadership style, language and signs, processes and practices and the definition of the success, and thus established the Organizational Culture Assessment Instrument (OCAI) based on the framework of competitive value put forward by them. OCAI divides the organizational culture into four types: Clan/Collaborate, Hierarchy/Control, Market/Compete and Adhocracy/Create according to the main pairing dimensions, which are the flexibility and self-determination versus stability and control and the internal focusing and integration versus external focusing and polarization. In Figure 3.1, the values represented on the upper left quadrant emphasize the interior and focus on organization and the values represented on the right lower quadrant emphasize exterior and control. Similarly, the right upper quadrant emphasizes exterior and organization and the left lower quadrant focuses the interior and control.

These four types are correspondingly judged from each four measuring items among the content of the six items of the organizational culture.

[11]Ma Mingfeng. Ways and Tools of the Organizational Culture Assessment [J]. Jiangsu Commercial Forum, 2007(2): 120–122.

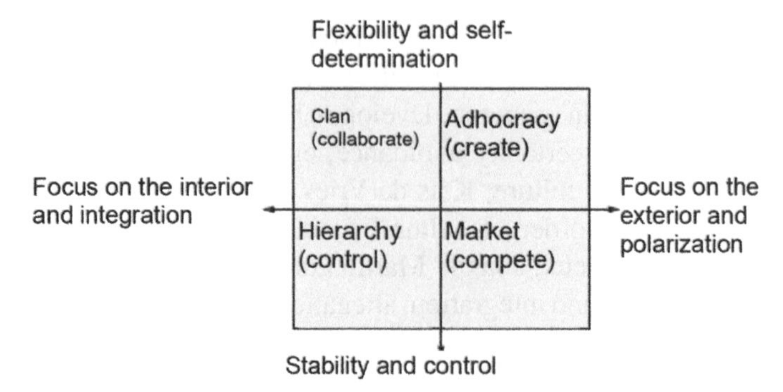

Figure 3.1 Cameron and Quinn's cultural assessment model about competitive values.

The content of the six items are dominant characteristics, leadership style, staff management, organizational cohesion, strategic focus and successful speculation. Among them, the dominant characteristics and leadership style belong to the basic assumption dimension; staff management and organizational cohesion belong to the interaction model dimension; strategic focus and successful speculation belong to the organizational direction dimension. Each item involves four choices and each choice is the description of some kind of the organizational culture. Respondents need to distribute the score of 100 points to each option according to the description of each option and the degree of fitness to the organizational culture. For example, for the content of the first item, if option A extraordinary fits the organization, option B and C fit a little and option D doesn't fit at all then you can give option A 55 points, 20 points each for option B and C and 5 points for option D. Just make sure the total score of the all options is always 100 points. In addition, OCAI measures both the current culture of the organization and the culture that can be reached by the organization within 5 years. Cameron and Quinn believe OCAI is very helpful in identifying the type, strength and consistency of organizational culture.

3.3.1.2 Daniel denison model (OCQ)

Daniel Denison believes that organizational culture is a deep structure of the organization. It roots in the values, belief and assumption held by the organization members. Afterwards, Daniel Denison studied the individual cases of five organizations and he thought four kinds of cultural traits are significantly correlated with the organizational effectiveness. These four traits are mission, adaptability, participation and consistency. From the dimension of flexibility and stability, the culture of adaptability and participation of the

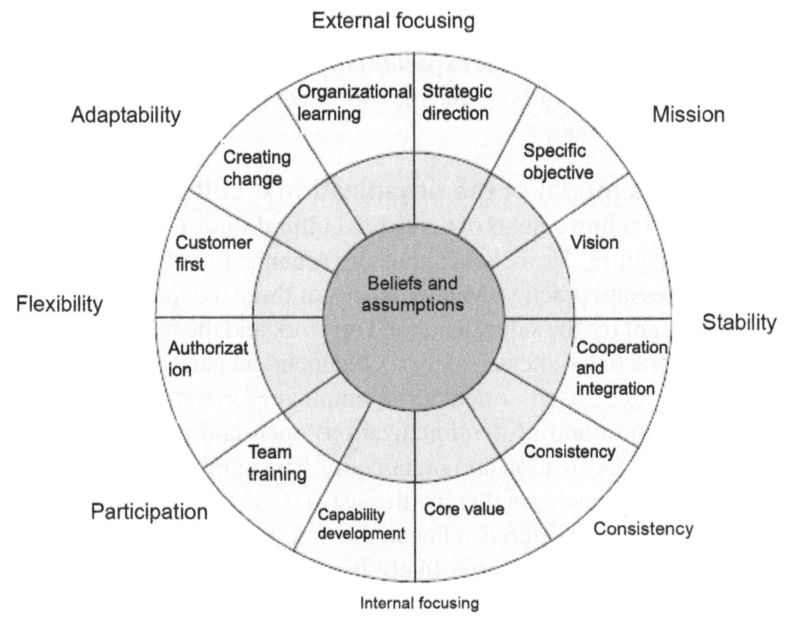

Figure 3.2 Daniel Denison's cultural assessment model (OCQ).

employees are much flexible and the culture of mission type and consistency are much stable. From the dimension of internal focusing and external focusing, the culture of adaptability and task-based have much bias toward the external focusing and the culture of the participation of the employees and the consistency have bias toward internal focusing.

The dimension of mission mainly measures the direction of the long-term development of the organization and it includes strategic direction, specific objectives and vision. The dimension of adaptability mainly measures the reaction of the organization while facing the external environment and it includes three aspects of creating change, customer first and organizational learning. Participation mainly inspects the construction of the sense of belonging and the allocation of responsibility of the organization and it includes authorization, team training and capability development. Consistency mainly examines whether the values of the organization members are in line with the core values of the organization and it also includes three aspects which respectively are the core values, consistency and cooperation and integration.

Daniel Denison worked out the Organizational Culture Questionnaire (OCQ) based on this and it includes 60 measuring items. OCQ is designed in accordance with the strong and weak performance of the organization not the division of the type of the organizational culture. It analyzes the differences between the existing culture and the expected culture so as to seek the foothold from the culture revolution through the measurement of the four typical traits of the organizational culture.

3.3.1.3 Hofstede's model of the organizational culture

Hofstede applied the achievements of his state's cultural study to the evaluation of organizational culture. He believes that the organizational culture is the complex of the vales and practice. Values consist of three independent dimensions: the requirement for the safety, centered on work and the requirement for the authority. The practice is measured by six independent pairing dimensions: process orientation and results orientation, employee orientation and work orientation, localization and professionalization, open and close, loose and strict control and standardization and practicality. The empirical study shows that the difference is much smaller in the aspects of the values for each organization and the main difference lies in the practical part. The difference in the aspect of values may be more likely to appear among cross-country, cross-nation and trans-regional researches.

3.3.1.4 O'Reilly, Chatman and Caldwell's organizational culture profile (OCP)

O'Reilly, Chatman and Caldwell believe that organizational culture is the system of values shared by all the organization members. The evaluation of the organizational culture shall focus on the differences between the individual values of the organization members and the values shared by the organization, namely the degree of fitness between members and organization. For it just the degree of fitness affects the outcome variable of the individual, such as the organizational commitment and turnover intention etc. Based on this consideration, they compiled Organizational Culture Profile (OCP) to measure the fitness between the people in the organization and culture. Organizational Culture Profile measures the 7 dimensions, namely innovation, stability, respect for employees, results orientation, attention to details, team orientation and aggressiveness, of the enterprise culture through the measurement of 54 projects. In consideration of the significant influence on the organizational culture of the senior management, the respondents of the OCP profile are senior management of the organization. Besides, it adopts a kind of scoring method, the Q-sorts. This method requires dividing the measuring items into nine types according to the expected most to the expected least or qualified most to qualified least. The number of the items included in the types is distributed as 2-4-6-9-12-9-6-4-2, which actually is a filling in method of the ipsative kind. While filling answers, it requires the professionals to assist alongside. For the usage limit of the OCP profile, the upcoming researchers developed this and thus created the various versions of the OCP profile.

3.3.1.5 CAT culture assessment tool

American company NEW LWADERS comes up with CAT (Culture Assessment Tool) including corporate culture assessment (CAT-I) and cross-cultural communication assessment (CAT-II) two systems to study cultural conflict and integration, but emphasizing on the comparison and analysis of cultural conflicts on the basis of modern cultural theory. Wherein, corporate culture assessment consists of 11 major cultural categories, 180 statements of 21 segmentation indexes, which can be used to confirm 39 possible cultural directions. And it is a system, on basis of Internet, named fill in the blanks with answers that participants need to answer their degrees of agreement. In addition, it is an interactively dynamic system which can automatically request participants to give their advices and comments on these questions that attract more people's attentions, and in this way, we can collect some details of some certain culture.

3.3.1.6 Goffee & Jones's Double S cube

Goffee & Jones believe that organization's characteristics can be explained by identifying its sociability and solidarity. Solidarity means the similarity of member's mode of thinking while sociability means the degree of member's mutual respect and care. Corporate culture can be divided into four parts:

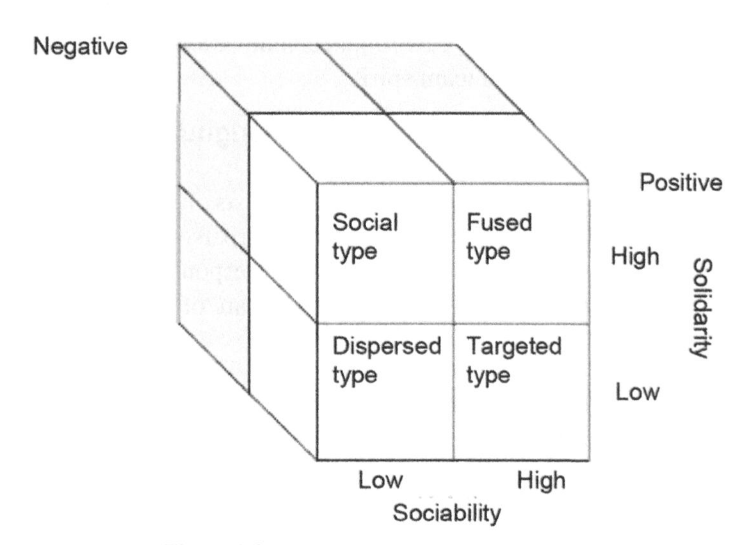

Figure 3.3 Goffee & Jones's double S cube.

social, fused, dispersed, targeted, and each type has positive and negative effects according to those two dimensions.

Research tools of this model include four types: (1) observation scale, involving the observation of entity space, communication, time management and self-identity; (2) questionnaire of corporate culture, including 23 questions, used to judge the type of corporate according to the score of organization's solidarity and sociability; (3) questionnaire of culture type, there being respective 6 items under four different types of organizational culture, which 24 items in total, used to judge whether the culture type belongs to the negative side or the positive side; and (4) affirmation of the culture type by analyzing key issues.

3.3.1.7 Cheng, B. S.'s VOCS

Cheng, B. S. believes that previous organizational culture measurement study at the individual level lacks relevant theoretical framework. Therefore, he thinks that organizational culture shall consist of three parts: basic assumption, external value and artificial decorations according to Schein's cultural theory. And he goes much further, thinking that basic assumption has 5 dimensions, namely the relationship between organization and environment, fact and truth, and the essence of human nature, human activity, and human relationship, and on this basis he designs VOCS (Values in organizational Culture Scale) to measure corporate culture. And it turns out that there are two higher dimensions and external adaptation values include social responsibility, honesty and amity, customer orientation, and scientific truth seeking, while the internal integration values include sincerity, performance, innovation, sharing the joys and sorrows with each other, and team spirit.

3.3.1.8 Corporate culture assessment of Guanghua school of management

According to the results of case analysis, its scale consists of 7 dimensions and 34 quizzes. And the 7 dimensions respectively are interpersonal concordance, impartial bonus-malus, normative integration, social responsibility, customer orientation, and daring to innovate as well as taking care of the development of employees.

3.3.1.9 Corporate culture assessment of Tsinghua SEM

Its scale consists of 8 dimensions and more than 40 quizzes. And the 8 dimensions respectively are customer orientation, long-term orientation, result orientation, behavior orientation, control orientation, innovation orientation, harmony orientation, and employee orientation.

3.3.1.10 Scale of CCMC

CCMC, taking OCAI as a reference, divides the evaluation of corporate culture into 6 parts: the types of corporate culture, concept orientation, core values, environment, leadership skill, personal value and occupation orientation, which all constitute the local culture assessment scale.

Here taking the L-PCAI (Leader's Preferred Culture Assessment Instrument) of a consulting company in China as an example, which developed on the basis of OCAI scale.

Leader's Preferred Culture Assessment Instrument (L-PCAI)

L-PCAI estimates a leader's enterprise culture tendency according to 6 aspects: management characteristics, organizational leadership, staff management, organizational cohesion, strategic objectives and success criteria. There are

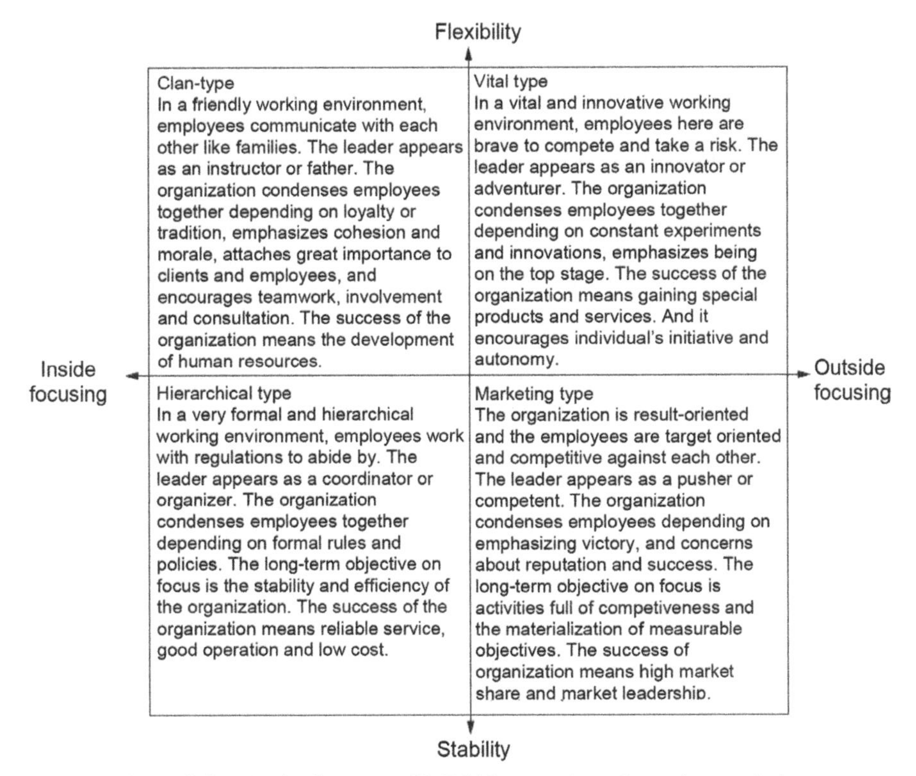

Figure 3.4 Matrix diagrams of L-PCAI enterprise culture characteristics.

6 dimensions and 24 test items. And each dimension is followed by 4 declarative sentences corresponding to 4 types of enterprise culture. For a certain enterprise, its enterprise culture, at any point in time, is a mixture of 4 enterprise culture, namely clan-type, vital type, hierarchical type and marketing type.

The respondents of L-PCAI Questionnaire are high-level leaders of enterprises. Each question involves 4 declarative sentences with a total score of 100 points. Respondents are required to fill in the blank with a score corresponding to the current situation and the future expectation in the back of each corresponding declarative sentence. The higher the score is, the more confirming the situation is. 0 or 100 are allowed for each question. For example, the scores of 4 questions could either be: 30, 54, 0, 16, or be: 100, 0, 0, 0, but the total score should be 100 points.

Note: In the questionnaire, the column "present" means the current reality of an enterprise, and the column "future" means the expected situation in the 5 years to come.

L-PCAI Questionnaire

Management Characteristics	Present	Future
**Your company is full of vitality and enterprise, and employees are eager to receive and bear risks.	B	
**Your company has a clear organization structure and a complete control system. Employees work abiding by rules and regulations completely.	D	
**Your company attaches great importance to the fulfillment and result of work, and the employees also value competition and achievement a lot.	C	
**Your company is like a big family where there is space for individuality, and employees can share joys and sorrows.	A	
Total points	100	100

Organizational Leadership	Present	Future
**The leader of your company is employees' instructor, caregiver or promoter.	A	
**The leader of your company is an entrepreneur, innovation promoter or reformer.	B	
**The leader of your company is a coordinator, organizer and a person improving operating efficiency of your company.	D	
**The leader of your company is pragmatic, energetic and only valuing the work result.	C	
Total points	100	100

Staff Management	Present	Future
**The management of your company is characterized by hyper-competition, high requirements and high achievements.	C	
**The management of your company is mainly dominated by teamwork, participative management and gaining the common ground.	A	
**Your company seeks for stability of employment and relationship between employees, and consistency and predictability of employees' actions.	D	
**The management of your company is full of personal adventurism, freedom, innovation and uniqueness.	B	
Total points	100	100

Organizational Cohesion	Present	Future
**Your company's cohesion comes from formal regulations and policies. And it is very important to keep the company run stably.	D	
**Your company's cohesion is formed by objective accomplishment and fulfillment valuing. And making progress and win ar the theme of your company.	C	
**Your company's cohesion comes from valuing innovation and development. And your company focuses on eliminating boundary so as to integrate everyone into the whole	B	
**Loyalty and mutual reliability are the sources of company's cohesion. And it is very important that employees bear their own responsibility.	A	
Total points	100	100

Strategic Objectives	Present	Future
**Your company values individuals' development, reliability, openness and continuous involvement.	A	
**Your company values sustainability and stability, and emphasizes efficiency, control and smooth running.	D	
**Your company values gaining new resources and creating new challenges, and encourages attempting new things for opportunities.	B	
**Your company emphasizes competitive actions and fulfillments. And reaching objectives and winning in market are of the most importance.	C	
Total points	100	100

Success Criteria	Present	Future
**Success refers to your company's victory in market, exceeding competitors and being the leader in market competition.	C	
**Success means that your company possesses the newest or special technologies or services. And the company is the leader and innovator of these technologies and services.	B	
**Your company gains a success only on the basis of employees' development, teamwork, employees' promise and paying close attention to employees.	A	
**Efficiency is the foundation of your company's success, and the key is reliable delivery, smooth plans and low cost.	D	
Total points	100	100

Scoring volume

Current status of culture

Clan-Type	Vital Type	Marketing Type	Hierarchical Type
1A	1B	1C	1D
2A	2B	2C	2D
3A	3B	3C	3D
4A	4B	4C	4D
5A	5B	5C	5D
6A	6B	6C	6D
Sub-total A	Sub-total B	Sub-total C	Sub-total D
Average	Average	Average	Average

Preference of culture

Clan-Type	Vital Type	Marketing Type	Hierarchical Type
1A	1B	1C	1D
2A	2B	2C	2D
3A	3B	3C	3D
4A	4B	4C	4D
5A	5B	5C	5D
6A	6B	6C	6D
Sub-total A	Sub-total B	Sub-total C	Sub-total D
Average	Average	Average	Average

Type graphs of cultural assessment

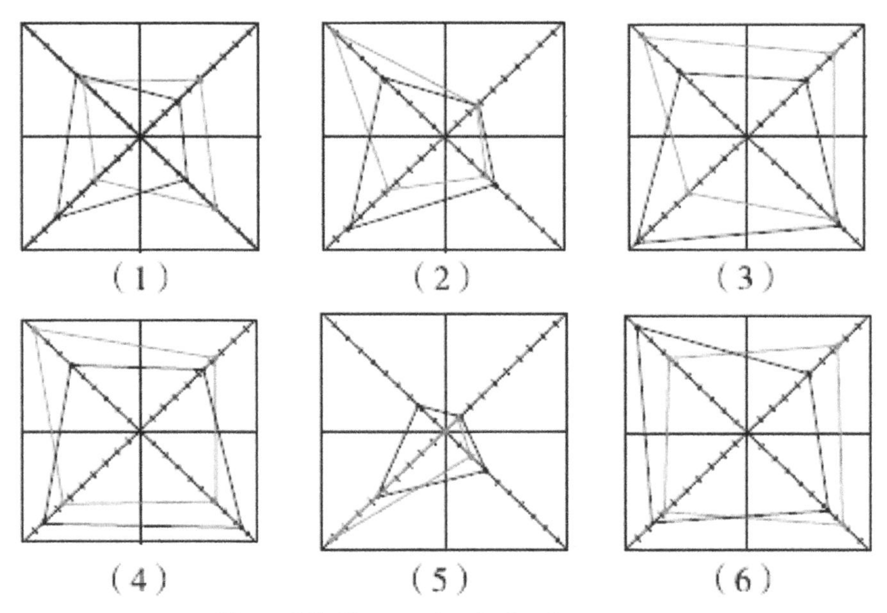

Figure 3.5 Type graphs of cultural assessment.

This model, according to the inside focusing and outside focusing, flexibility and stability, divides organizational culture into four types, namely clan-type (upper left), vital type (bottom left), hierarchical type (upper right) and marketing type (bottom right). The result of organizational culture assessment is shown with lines in the center of graphs. Wherein, black lines represent the current organizational culture, whereas grey lines represent organizational culture expected by members in the organization. The nearer to the edge of quadrant, the more outstanding the organization's performance is in this kind of cultural type. Take Figure 3.2 as an example. It possesses strong hierarchical organizational culture. However, that is what the members in clan-type want to achieve. Every graph in Figure 3.5 are all practical results in different organizations.

One of the reasons why there are various dimensions and tools in studying organizational culture is that organizational culture is extensive and inclusive in scale. It includes a series of complex and mutual related factors. As a result, it is impossible to include all related factors in the process of the diagnosis and assessment of organizational culture. In the sense, it is so crucial for the study of organizational culture to choose a basic theoretical framework or model in order to focus on the contents in study.

3.3.2 Models and Tools of School Culture Assessment

Models of school culture assessment include school culture type assessment model and school culture outline assessment model. Atmosphere assessments such as open-control assessment model, healthy-morbid assessment model and so on belong to the former.

3.3.2.1 Open-control school culture atmosphere assessment model

Open-control school culture atmosphere assessment model is a research achievement of A. W. Halpin and D. B. Croft on primary school culture. They designed *The Organizational Climate* with totally 64 items of assessment indicators divided into eight dimensions among which four dimensions (barriers, intimacy, doing things carelessly and mental state) are used to analyze the headmaster's behavior characteristics and the rest four dimensions (paying more attentions to work, coldness, thoughtful concern and propulsion) are used to analyze the teaching faculty's behavior characteristics. By inspecting different types of interactions among teachers and between teachers and the headmaster and in accordance with the sequence from open to close, six different kinds of school atmosphere can be set up: open, autonomous, control, non-governing, parental and close.

Open school culture atmosphere: typical characteristics of this atmosphere are high trust and low loose organization. Both the headmaster's and teachers' behaviors are full of passions. The headmaster thoughtfully concerns about teachers and provides supports for their jobs. Teachers cooperate with each other happily and can devote themselves to completing works. The headmaster's leadership behaviors are simple to conduct and there is no need to strictly supervise teachers or to act strictly in accordance with the rules.

Independent school atmosphere: typical characteristics of this atmosphere are high coldness and low concern and the whole school organization is almost completely free. In such a school organization, "mental state" and "intimacy" is relatively high while there are fewer "doing things carelessly" and "barriers". Teachers work according to their desires in order to satisfy their own social needs. On the other hand, the headmaster is relatively indifferent, relatively loose in supervision and the extent of thoughtful concern to teachers is relatively low.

Control-type school atmosphere: typical characteristics of this atmosphere are paying highly attention on work, high barriers and low intimacy. Such a school organization is usually at the cost of sacrificing social life in order to work hard. Teachers' non-teaching work is heavy and interpersonal communications are relatively few. The headmaster himself or herself works hard and plays a dominant role in the process of school management but he or she is cold and doesn't care much about the subordinates.

Laissez-faire school atmosphere: typical characteristics of this atmosphere are paying low attention on work, low barriers and high intimacy. In such an organization, the headmaster highly cares about and considers for the staff and is warm but doesn't care about work. As a result, teachers care more about personal life but are far away from work. Its result is that the organizational atmosphere is very friendly but the effectiveness of work is very slight. That is to say, good organizational atmosphere is at the cost of work efficiency.

Patriarchal school atmosphere: typical characteristics of this atmosphere are low coldness and high consideration.

In such an organization, the headmaster is considerate and warm but the emphasis on work is overly abrupt and thus effectiveness of work is extremely low; and teachers have no overly heavy burdens but they usually can't get along very well with each other but form different competing factions. Headmasters in such schools are usually deemed as relatively merciful dictators.

Close-end school atmosphere: typical characteristics of this atmosphere are low trust and loose organization. The headmaster and teachers seem to complete tasks. The headmaster emphasizes insignificant daily trifles and teachers respond bluntly with several mutterings. The headmaster lacks caring about teachers, strictly supervises teachers, handle affairs completely in accordance with the rules and leader effects are not good. Teachers often feel frustrated and behave indifferently in activities.

Enlightened by *The Organizational Climate*, successive scholars revised *The Organizational Climate Descriptive Questionnaire* again on the basis of criticizing its internal logic results, separately formulated primary school version and middle school version, and measured four kinds of school culture atmosphere.

Open school atmosphere: the headmaster and teachers respect and cooperate with each other. The headmaster is willing to listen attentively to teachers' suggestions, praises teachers in good time and usually, respect teachers' career development and protects them from interference of heavy affairs to the best of his or her ability. In the same way, teachers can communicate teaching experience with each other, sincerely cooperate with each other and are willing to struggle for the teaching of schools.

Instruction-based school atmosphere: on one hand, the headmaster extremely does not respect teachers' career development and also does not consider teachers' individual demands, but he or she overly reinforces his or her own control for teachers and at the same time gives teachers overly heavy workloads. However, on the other hand, no matter how the headmaster requires, teachers only concentrate on their own education and teaching. Colleagues cooperate with each other and their personal friendship is also good.

Loose school atmosphere: the headmaster is willing to listen attentively to teachers' suggestions, supports teachers' professional development, pays attention to reducing teachers' workloads, and gives teachers rights for free development. But teachers are not willing to accept the headmasters' behaviors. Not only they don't like the headmaster but also colleagues hardly communicate with each other, isolate with each other and disrespect each other. They do their own things and almost devote nothing to their school.

Close school atmosphere: the headmaster and teachers are only completing tasks and there is no respect with each other at all. The headmaster is busy issuing administrative orders and imposing workloads on teachers. Teachers respond bluntly. Teachers are indifferent with each other, and they don't care about each other nor respect each other.

3.3.2.2 Healthy-morbid school atmosphere assessment model

Through analyzing the healthiness of interpersonal relationships between students, teachers, managers, and communities, the model assesses whether the school culture is healthy or not.

Based on Talcott Parsons' sociological theory framework, this analysis framework designs *Organizational Health Inventory* from three levels, namely technology, management and system, to assess school atmosphere. And there are totally three versions for primary school, junior middle school and senior high school respectively.

The model points out that the characteristics of healthy schools are that students, teachers and headmasters are consistent in behaviors and jointly work hard for the same teaching success. System of such school is highly perfect, wherein teachers won't be confused unreasonably and viciously by outside forces and the headmaster provides all kinds of resources required in work for teachers, being able to have an effect on superiors and providing a variety of support for teachers of the school to the best of his or her ability. Teachers like their colleagues, jobs and their students. They are motivated by the pursuit of excellence in academics. Teachers are very self-confident and also believe in their students. As a result, they establish relatively high and achievable goals. The study environment is ordered in which students work hard showing respect for students who are better in studies. The headmaster's behaviors are friendly, open, equal, and full of constructiveness. Such headmasters expect to get the best working behaviors from teachers.

Morbid school atmosphere appears very weak in front of outside collapsing force. Teachers and headmasters are easily attacked by unreasonable parents' demands and schools are tortured by all kinds of unreasonable demands from the public. Such school is short of a capable headmaster. The current headmaster considers less about directions and structures of the school and encourages teachers' little, and his or her effect on the superiors is not worth mentioning. Teachers do not like their colleagues and jobs and they behave indifferently, suspiciously and defensively. They cannot get teaching materials, supplies and supplementary goods and materials timely when required. Pressure for academic excellence is very small and both teachers and the headmaster trifle with students. In fact, hard-working students who want to study well even be laughed at by their companions and will also be deemed as threats by teachers. As a result, each person seems to "kill the time".

Characteristics of these healthy and morbid school atmospheres are at two extremes, and most schools are sitting in the middle. Situations of school culture atmosphere can be known by examining the degree of conformities or disconformities of the two mentioned states.

3.3.2.3 Regulatory orientation—human orientation school climate assessment

This model is theoretically based on the fact that school culture contains two ultimately opposed cultures—adult social culture represented by teachers and peer group culture represented by students, which two are in conflict. So, school culture climate can be reflected from the control of the school over students.

Regulatory orientation school climate: for school, the top priority is to maintain order. Students' appearances and behaviors and parents' social status are unchanging. Teachers regard school as a specific organization, in which the classes of teachers and students are entirely different, the form of right and information communication is unidirectional and descending, and students must take the order of teachers absolutely. Teachers refuse to understand students' behaviors, and treat their naughtiness as a kind of offend. They think students are unruly and don't have a sense of responsibility. So, they have to adopt punishment as the approach of control. Inexorability, satire, suspicion and distrust are the primary features of regulatory orientation climate.

Human orientation school climate: school observes students' studies and behaviors from aspects of psychology and sociology. Self-discipline has taken the place of teachers' strict control. Then the atmosphere is democratic, the contact between teachers and students is free and the communication is bidirectional. The ability of students on making their own decisions is improved, too. Such school climate not only put stress on individual importance but also emphasizes the importance of creating a kind of school atmosphere that can meet students' need.

Through the three models mentioned above, the pointcut of school climate assessment is interpersonal relationship. Open—control type mainly assesses the relationship between principal and teachers, healthy—morbid type mainly stresses the relationship between all subjects in school and community outside, and regulatory orientation—human orientation mainly emphasizes on the relationship between teachers and students. All of these patterns can describe the state of school culture, but they usually focus on one aspect and cannot show the entire outline of it.

3.3.2.4 Assessment models of school culture outline[12]

Assessment models of school culture outline can give a comprehensive assessment on school culture, and to much degree make up for the deficiency of assessment models of school climate. Based on the theoretical framework of competing values, these models, after concluding from many influencing factors on organization efficiency, work out two groups of opposite standards: stability and control—flexibility and adaptability, with emphasis on internal management and integration and focus on outer competition and differences. Besides, it also creates four main culture types (see Figure 3.6): hierarchical type, market first type, clan type and temporary organization type.

Hierarchical type: it focuses on internal stability, stresses systemization and institutionalization, possesses a clear decision-making frame, a standardizing system and strict hierarchy, and requires to definitude responsibilities

[12]Cameron and Quinn. Organizational Culture Diagnosis and Reform [M]. Xie Xiaolong, translation. Beijing: China Renmin University Press, 2006: 27–36.

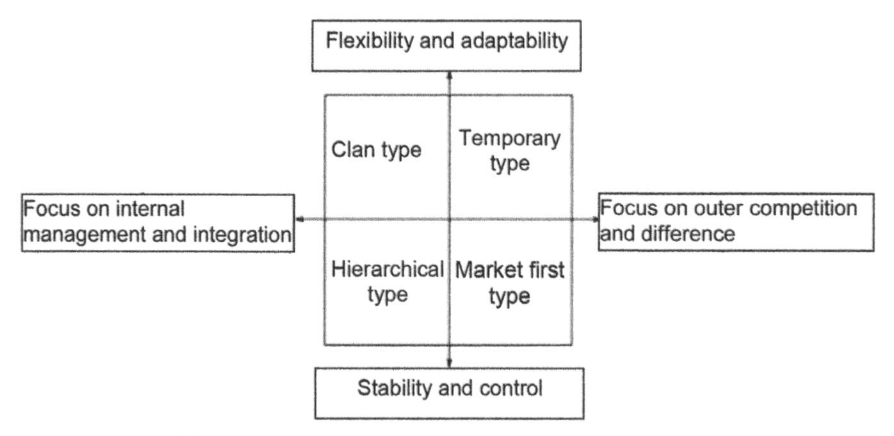

Figure 3.6 Quadrant figure based on competing values.

and obligations. The leader shall possess excellent ability of organization and coordination, and put much emphasis on internal management and control.

Market first type: it focuses on external affairs. For this type of culture, competition and productivity are the core values, and achievement is highly stressed; the aim of managers is to improve the productivity, competition and profit. For them, surpassing their opponents and being the market dominate is the most important index.

Clan type: it focuses on staff and its working environment is humanized. So, all the staff will be passionate in work, willing to devote them to work, always united as one, love to help each other and like to share the common values and aims. The leader of this type is more like an instructor or parent.

Temporary system type: it focuses on the future and hammers at developing new products and services to adapt to the changes. There is no clear organizational structure and staff's responsibility is based on the task. The leader of this type shall be creative and dare to take risks.

Cameron and Quinn, the creators of these models, also design a questionnaire about organizational culture assessment to conduct the assessment of organizational culture. This questionnaire measures from 6 aspects, namely main features, leadership, staff management, adhesive forces of organization, strategic focuses and standards of success. There are four choices of each aspect, and the participants need to allocate 100 scores to these four choices by using scoring method. This approach can reveal not only the current status but also the expected status of the organization.

Organizational culture assessment questionnaires in the models of culture outline assessment[13]

1. Main Features		The Current	The Expected
A	Organization is a place filled with humanization; it's the extension of family and people here are united as one.		
B	Organization is filled with flexibility and entrepreneurship; people here dare to adventure and bear responsibility.		
C	Organization is utilitarian; people here are aimed at completing task. Staff possesses excellent ability and expects success.		
D	Organization is controlled strictly and has clear regulations; people here shall comply with the regulations.		
	Total points	100	100
2. Leadership of the Organization		**The Current**	**The Expected**
A	The leader of organization is always commented as an instructor, promoter or cultivator.		
B	The main leadership style of the organization is entrepreneurship, innovative and adventurous.		
C	The main leadership feature of the organization is "no nonsense" and filled with ambitions and strong utilities.		
D	The main leadership feature of the organization is clear, organized, smooth and efficient.		
	Total points	100	100
3. Staff Management		**The Current**	**The Expected**
A	The management features are group working, minorities submitting to majorities, and with good participation degree.		
B	The management focuses on individualism, adventure, innovation, freedom and self-realization.		
C	The management style shows strong competition and strict requirements and standards.		
D	The management style is to confirm the employment relationship, and the relationship is foreseeable, stable and coincident.		
	Total points	100	100

[13]In original questionnaires, "The current" and "The expected" are subordinate to two questionnaires. Here, to save space, they are merged into one questionnaire.

	4. Adhesive Forces of Organization	The Current	The Expected
A	The beliefs of organization are loyalty and mutual trust. People have the sense of responsibility to bear obligations.		
B	People are united for innovation and development, and their aim is to be the pioneer.		
C	People share the common thoughts, namely success and completion of plan. Being ambitious and to be successful are the common goal.		
D	The regulation and policy are standard, and it's very important to maintain organization run smoothly.		
	Total points	100	100
	5. Strategic Focuses	Status	Expecting Status
A	The organization stress the development of human resources, mutual trust, frank and straight communication, and staff's participation.		
B	The organization is mainly engaged in exploring new resources and facing new challenges, being curious about new things and seeking for opportunities are the staff's value.		
C	The organization aims at being competitive and successful. Its main strategy is striking its opponents and being the winner.		
D	The organization expects to be persistent and stable and it emphasizes on efficiency, control and smooth operation.		
	Total points	100	100
	6. Standards of Success	The Current	The Expected
A	The success for organization is success on human resources, group working, staff's contribution and staff care.		
B	The success for organization is that it has special and new products, and organization is the leader and creator of products.		
C	The success for organization is occupying market share, beating opponents and being the leader of market.		
D	The organization regards efficiency as the basis of success. It emphasizes on tray-passed and steady working plan and decreasing cost.		
	Total points	100	100

Organizational culture outline can be painted out based on the results of the questionnaires. The specific steps are as follows. First, add up all the scores in "The current", and divide by 6, the result is the average score of option A. Similarly, calculate out the average scores of option B, C and D. Then, do the mathematics in "The expected" with the same method. The scores of option A, B, C and D are respectively the final score of clan type culture, temporary type culture, market first type culture and hierarchical type culture.

Next, draw the organizational culture status outline. Add the scores of the four options to the corresponding quadrant diagonals in the outline (see Figure 3.7). For example, write down the scores on the scale of top left quadrant. Use straight line to connect the scores in the outline, and the quadrangle is the profile of culture status.

Finally, draw "The expected" figure of organizational culture. The drawing method and steps are the same as mentioned above except for that change the straight connecting line to imaginary line for distinguish. Then the organizational culture outline is drawn. Figure 3.8 is the example of completed expected organizational culture outline.

Multiple kinds of analyses can be made with the outline drawing of enterprise culture. In this figure, we can conclude its dominant culture type

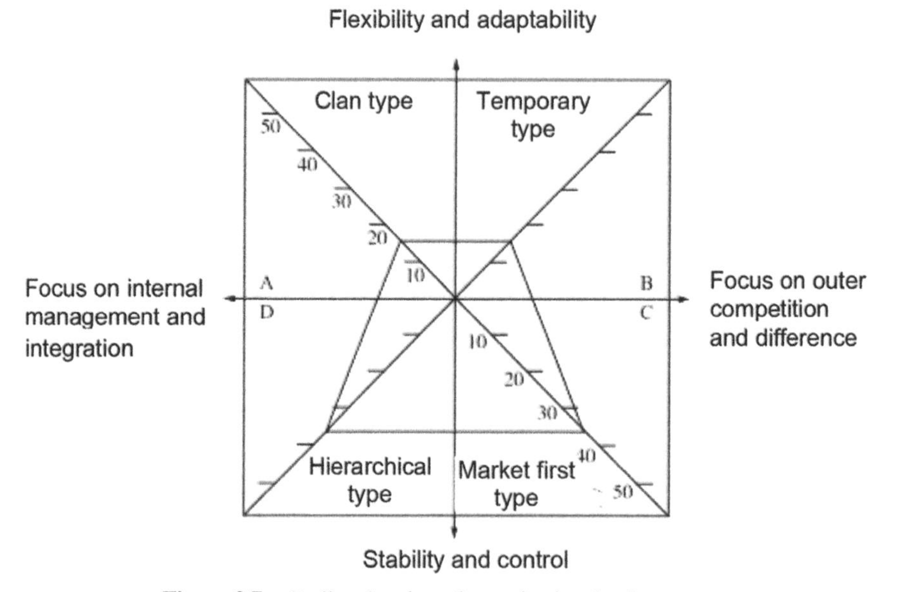

Figure 3.7 Outline drawing of organizational culture status.

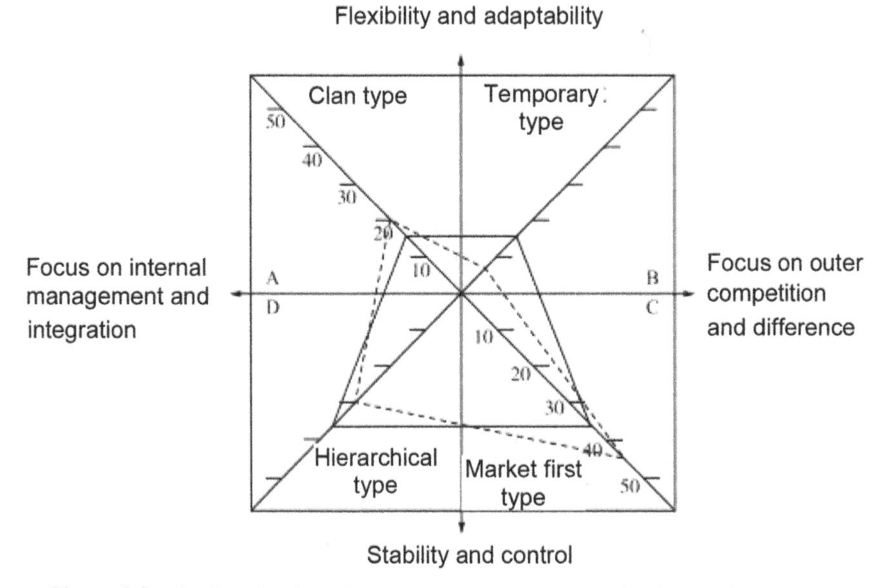

Figure 3.8 Outline drawing of the current and expect organizational culture values.

through the highest scored quadrant, and we can see the disparity between the current and expected status and then determine the reform direction. The score of culture outline decides the culture strength, and from this figure we can see the culture strength of the organization. In addition, the coincidence among different departments and different groups is also can be deduced. The 6 cores of organizational culture, namely main features, leadership of organization, management of staff, adhesive force of organization, strategy focus and standards of success can also use six figures to compare with organizational culture outline. The more coincident between them, the less difference there is. This figure can be used to conduct comparison with other organizations within the industry too.

Cameron and Quinn are always engaged in improving this model and exploring other supportive tools to perfect it. This model has been widely used in various organizations, including schools. Practice shows that it is efficient and accurate.

3.3.3 Assessment Models and Tools of Project Practice

Most of the above-mentioned models and tools of school cultural assessment are coming from abroad, and they shall be tested before applied at home. At present, the school cultural assessment at home is getting matured and there are many models and tools of school culture assessment emerging. On the basis of above research, I lead the project team to design *Assessment Questionnaire of the School Culture Development State* based on the driven model of school culture. This questionnaire is mainly achieved by Dr. Xu Zhiyong through taking reference to the existing research results and through practice inspection. It is of favorable reliability and validity.

3.3.3.1 Tool design process

The design of questionnaire and scale is based on the following fundamental assumptions. Principal's school-running ideas, values and management behaviors do have a direct influence on the educational performance, but they can only come into effect through teachers, the intermediary. Teachers' inner and outer satisfaction degree is influenced by their identification with the school-running ideas and values of the principal and the management behaviors of the school. The two variables, teacher's inner and outer satisfaction degrees will directly influence the educational performance of the school while the educational performance of school is directly affected by the external environment. And there are 7 dimensions on the basis of above assumption: identification with the school-running ideas, values of school leaders, school management behaviors, teachers' inner satisfaction degree, teachers' outer satisfaction degree, educational performance of school, and the external environment of school.

The subject design of school leaders' values makes reference to the maturity scale of CEO Values measurement index of some people including Bradley R. Agle. It also includes 3 effective test items: "most executives are very caring and compassionate", "executives are advocates of good deeds", and "most executives are good at thinking for others".

Some subjects of school management behaviors make reference to the measurement index of man-oriented organizational culture instrument of Jennifer A. Chatman & Karen A. Jehn. And there are 4 effective test items: "company respects personal rights of its staff", "company treats its staff fairly", "company supports its staff sufficiently", and "executives often listen to its staff and know their needs".

The design scale of teachers' inner and outer satisfaction degrees makes reference to Chen Zhizhong's maturity scale of "social support and work satisfactory of teachers". This scale divides satisfaction degree into two dimensions: inner and outer satisfaction degrees and the query language of

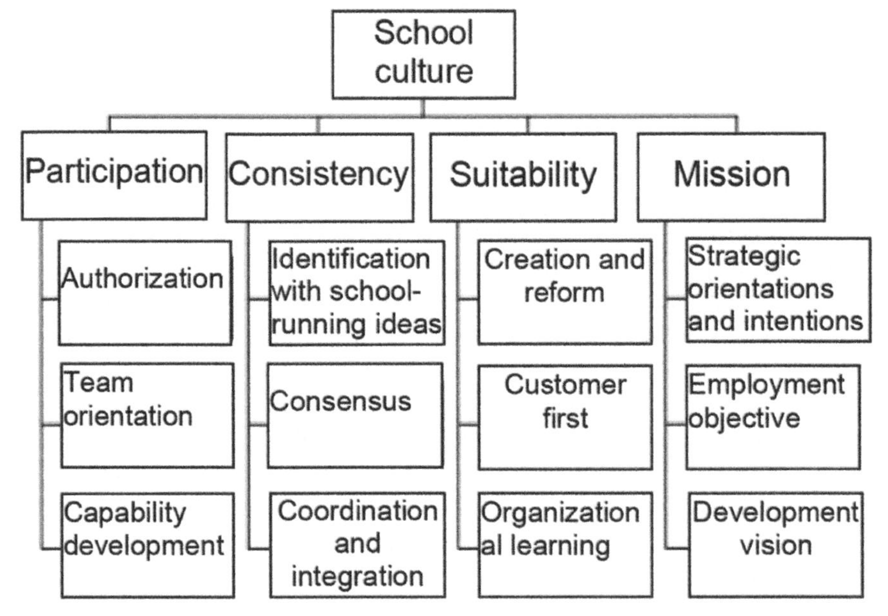

Figure 3.9 Structure of the assessment questionnaire of the development state of school culture.

the scale have a strong flavor of Taiwan accent, which is adjusted according to reality of cultural construction while designing.

The subject of identifying with school-running ideas, educational performance of school, and external environment of school are self-designed. And its credibility and effectiveness need to be tested.

We have picked out 230 teachers from 4 primary schools for the test after completing the first draft of the questionnaire. And after the test, we have picked out 20 teachers to be interviewed about the subject and presentation of questionnaire. Then, we did the item analysis, differential degree of inspection, reliability analysis and validity analysis on the basis of test data and did some modifications and perfection of questionnaire according to the interview results.

3.3.3.2 Content structure of questionnaire

Questionnaire consists of 4 first-class dimensions, 12 second-class dimensions and 40 questions (details in the Figure 3.9).

Questionnaire is in the form of Likert Scale: 1 point means that teachers think that the description of the relevant item "totally unfits to" the actual situation of this school while 5 point means the description of relevant item "totally fits to" the actual situation of this school. And there must be only one answer to each item. This questionnaire mainly consists of the following 4 dimensions.

The participation of school culture: it mainly inspects and measures teachers' work capability, ownership and responsibility. 3 second-class dimensions respectively inspect whether the authorized staff are really authorized and undertakes the responsibility, staff's ownership and their work motivation; team orientation inspects whether school encourages teacher to have mutual cooperation and whether teacher achieves common goal by team work; capability development inspects whether school invests resources to cultivate teachers to promote their development and the satisfaction degree that school fulfills the desire of teachers' learning and development.

The consistency of school culture: it mainly inspects the interior cohesive force and centripetal force of school. And there are 3 second-class dimensions under the consistency: identification with school-running ideas inspects whether the school has the core values together believed in by the teachers and teachers' degree of identity toward school; consensus inspects whether the leaders of school are capable of making teachers reaching an agreement on the key issue; coordination and integration inspects whether the departments of school can cooperate in an effective way.

The suitability of school culture: it inspects the suitability of school against the external environment, mainly including the ability to capture and reaction velocity towards surrounding environment and parents' changes and needs. And there are 3 second-class dimensions under it: creating of innovation inspects whether school would observe the external environment carefully, thus it can take measures immediately; customer first inspects whether school can be able to have good communicate with parents to form a harmony relationship between school and family; organizational learning inspects whether school encourages creation and innovation and is good at capturing opportunity to learn.

The mission of school culture: it mainly inspects whether school has lofty aspiration and specific goal. And there are 3 second-class dimensions under it: strategic orientation and intentions inspects whether school has clear strategic target and does the utmost to achieve it when carrying out the work; employment objective inspects whether school makes objective which closely related to development vision and strategy and could make teacher strive for it; development vision inspects whether teacher understands and identifies with the ideal condition of school in the future and striving for it.

The four dimensions form two main contradictory bodies and they are also the main conflicts which needed to be balanced and solved by a construction of school culture. And the success or failure of construction of school culture depends on the settlement of those two contradictory bodies.

3.3.3.3 Assessment decision

The spectral theory of a construction of school culture divides school culture into three zones: threadiness, banding and continuation. If the average scores of the development state of the school culture is over 4.5 (including 4.5), then we can judge that the school is in the location of continuous band and it belongs to school with deep culture; if the school's average scores is 4 to 4.4, then the school is in the location of banding culture zone and its level of development is in middle culture zone; if the school's average scores is 3.5 to 3.9, then the school is in the location of the transitory stage from threadiness to banding culture zone and it belongs to school with weak culture; and the school culture with the average scores lower than 3.5 belongs to invisible light zone and the school is suffering cultural crisis.

3.3.3.4 Analysis of the credibility and validity of questionnaire

We adopt consistency of credibility to verify the credibility and validity of questionnaire. Generally speaking, if the reliability coefficient α can reach 0.6, then we think that the consistency of the measurement result is good; if coefficient α reaches 0.8 or higher, then it shows that the consistency of the measurement result is very good. According to the Table 3.5, the coefficient α of *Assessment Questionnaire of the State of School Culture* in this survey is 0.970, so the consistency of measurement result of this questionnaire is very good. Besides, the coefficient α of 4 second-class dimensions are respectively 0.874, 0.905, 0.842, 0.922, which are all over 0.8 and there are two of them even over 0.9; and only two coefficient α, the coefficient α of capability development and customer first, are lower than 0.6 while the rest of 12 coefficient α are all over 0.6, and all this shows that the credibility and validity of this questionnaire are very high.

We adopt two methods to analyze the subject of this questionnaire to judge the adequacy of subject. First, extreme groups test: we regard the subject which has no significant difference as the subject with low adequacy by independently testing the samples of the high score group and the low score group. Second, homogeneity test: we regard the subject whose the product moment correlation coefficient of comparison of questionnaire and scale's total points is lower than 0.30 or which has no significant

Table 3.5 Analysis of the credibility and validity of questionnaire

Dimension		Number of Subject	Reliability Coefficient	Standardization of Reliability Coefficient
First-class dimension	Participation	11	0.874	0.888
	Consistency	10	0.905	0.907
	Suitability	9	0.842	0.852
	Mission	10	0.922	0.931
Second-class dimension	Authorization	4	0.725	0.733
	Team orientation	4	0.771	0.771
	Capability development	3	0.516	0.567
	Identification with school-running ideas	4	0.847	0.851
	Consensus	3	0.777	0.778
	Coordination and integration	3	0.730	0.773
	Creation and reform	3	0.806	0.807
	Customer first	3	0.453	0.488
	Organizational learning	3	0.711	0.720
	Strategic orientations and intentions	3	0.686	0.733
	Employment objective	4	0.833	0.837
	Development vision	3	0.809	0.819
Questionnaire in total		40	0.967	0.970

standard as the subject with low adequacy by analyzing the product moment correlation. If there is a subject regarded as the subject with low adequacy in both tests, then we will cancel it.

The result of the item analysis of *Assessment Questionnaire of the Development State of School Culture* is shown in the Table 3.6, the result of comparison of extreme groups shows that differences of 40 subjects between extreme groups reach a significant level; and the result of homogeneity test shows that there is only one subject's product moment correlation coefficient compared with questionnaire is lower than 0.30, but compared with extreme groups, it reaches a significant level. So according to all the settings shown before, all the subjects in this questionnaire can be reserved.

Table 3.6 Analysis of the subject of questionnaire

Subject	Comparison of Extreme Groups (T Value)	Homogeneity Test (Subject Is Related to Total Point)	Remarks
Q1	−11.673**	0.592**	Reserved
Q2	−6.064**	0.305**	Reserved
Q3	−10.681**	0.575**	Reserved
Q4	−13.159**	0.684**	Reserved
Q5	−10.596**	0.670**	Reserved
Q6	−16.514**	0.708**	Reserved
Q7	−15.437**	0.753**	Reserved
Q8	−14.925**	0.640**	Reserved
Q9	−11.798**	0.631**	Reserved
Q10	−17.244**	0.770**	Reserved
Q11	−12.331**	0.660**	Reserved
Q12	−17.001**	0.724**	Reserved
Q13	−13.210**	0.671**	Reserved
Q14	−15.135**	0.729**	Reserved
Q15	−15.425**	0.729**	Reserved
Q16	−16.257**	0.747**	Reserved
Q17	−15.079**	0.664**	Reserved
Q18	−14.868**	0.775**	Reserved
Q19	−13.574**	0.741**	Reserved
Q20	−13.108**	0.740**	Reserved
Q21	−11.721**	0.621**	Reserved
Q22	−11.896**	0.715**	Reserved
Q23	−9.161**	0.566**	Reserved
Q24	−16.321**	0.771**	Reserved
Q25	−2.563*	0.211**	Reserved

(Continued)

Table 3.6 Continued

Subject	Comparison of Extreme Groups (T Value)	Homogeneity Test (Subject Is Related to Total Point)	Remarks
Q26	–19.358**	0.752**	Reserved
Q27	–21.412**	0.769**	Reserved
Q28	–11.921**	0.542**	Reserved
Q29	–12.008**	0.566**	Reserved
Q30	–16.145**	0.741**	Reserved
Q31	–20.143**	0.798**	Reserved
Q32	–20.163**	0.851**	Reserved
Q33	–10.899**	0.684**	Reserved
Q34	–12.844**	0.757**	Reserved
Q35	–8.762**	0.520**	Reserved
Q36	–18.083**	0.747**	Reserved
Q37	–14.562**	0.795**	Reserved
Q38	–16.323**	0.739**	Reserved
Q39	–18.084**	0.781**	Reserved
Q40	–16.324**	0.704**	Reserved

Note: **means marking on the level of 0.01 ($P < 0.01$); *means marking on the level of 0.05 ($P < 0.05$).

4

Formulation of School Culture Scheme

School culture scheme refers to the planning scheme of school culture development or construction formulated based on the assessment report of school culture development. It is called school culture scheme for short. This scheme is fulfilled by the cooperation of university's project team and project school whose common duty is to accomplish formulation of school culture scheme. Its formulation is an important element for school culture construction. If school culture is considered as a delicious water boiled fish, and different schools have different methods and creativity, school culture scheme is, of course, the secret recipe for the dish. As it is hard to cater for all tastes, to cook a delicious dish is of great difficulty, let alone making feasible school culture scheme with bright spots. The purpose of this chapter lies in through comprehending the author's own experience in managing projects and study, expounding basic contents associating with the topic – formulation of school culture scheme.

4.1 Connotation of School Culture Scheme

4.1.1 Definition of School Culture Scheme

4.1.1.1 School culture scheme

School culture scheme has various titles in practice such as school culture plan, school culture SIS development program, school identity system plan and school culture construction program. In academic world, its title has not been fixed, as all those titles are clear in meanings. Specifically speaking, culture construction as a part of school development planning, SDP emerged in the international society recently overlaps in sense with that discussed in this chapter. School development planning, with the essence of making school development an entry point, adopts some optimization programs testing relationship between organizations and environments so as to improve school's performance.[1]

4.1.1.2 Differences with work plan

What differences between school culture scheme and traditional school work plan? Table 4.1 shows the differences.[2] And what the one differs from the other most fundamentally is that the former's working entry point focuses on culture which promotes school's package development, whereas that of the latter is isolated or dispersed and even in chaos.

Through analyzing and learning from other scholars' documents, we believe that school culture scheme is a whole program keeping a firm grasp on school culture and formed from designing a scientific educational philosophy

Table 4.1 Differences between school culture scheme and school work plan

School Work Plan	School Culture Scheme
To value top-down processing	To value the combination of down-top processing and top-down processing
To be fulfilled by the few in school	Stakeholders (governments, communities, parents, school's leaders and teachers)
Stable, whole or partial	Dynamic, whole
Lack of supervision and feedback	To emphasize supervision and feedback
To emphasize hardware and software development	To centralize students' development and to pay more attention to software improvement

[1]Office of China/UK Southwest Basic Education Project. Guide for School Developing Planning [M]. Beijing: Education and Science Publishing House. 2009:16.

[2]The same as ①. Page 30–37.

system and planning an educational practice system. School culture scheme and school work plan should be coherent in spirit and mutual reliable and even completely coincident. In our project school, school culture scheme could be all transferred to a development program with the topic – culture.

4.1.2 Functions of School Culture Scheme

4.1.2.1 Connection

School culture scheme is a bond associating assessment and improvement, a bridge joining educational philosophy and practice, and also a culture passageway of school's past, present and future. It is the carrier of educational philosophy, the guidebook of educational practice, the summary of school's cultural tradition and cultural practice, and also the direction of school's development. School culture development cannot accomplish at one stroke, instead, requires constantly efforts generation after generation. Unfortunately, school culture scheme is hard to be inherited due to administrative changes among school's leaders. Once a school culture scheme is made and legalized to be a political text, it can be provided with institutional guarantee to some degree.

4.1.2.2 Guidance

Function of school culture scheme is to lead school members' actions or school culture development to the direction expected by school culture scheme, to deliver information including expectation for school's development and code of conduct, and to clearly inform of what and how a school does. As for teachers, School culture scheme is the general principle of imparting knowledge and educating people. As for students, it is a platform for their healthy growth. And as for schoolmaster, it is the blueprint of leadership and management.

4.1.2.3 Harmony

As education is not an isolated island and school is not a Land of Peach Blossom. In a certain sense, school environments are complex, closely relating to factors like social politics, economy, culture and population. In educational practice system, school culture scheme has to take the introduction and application of community resources, parents' resources and social capital into consideration. Hence it is necessary to establish, coordinate and maintain the relationship between communities, parents and schoolmates including in scheme formulation stage. The school invites these stakeholders sincerely for the involvement.

4.2 Formulation of School Culture Scheme

4.2.1 Formulation Subject

School culture construction is a process with multi-dimensional interactions of multiple subjects. All the relevant stakeholders of school development are likely to become the schemers, facilitators or the destroyers, participators or the evaders of the school development and reform. Their behaviors and attitudes and daily practices are included in the school reform, so they will be assimilated and accommodated by the school reform and become one of the formations that cannot be neglected, as they may have a great influence on the school reform. In the model of trilateral cooperation and the stimulation of school culture, we divide the formulation subject of school culture into two parts: the key subject and the secondary subject, which can also be called the internal subject and external subject.

4.2.1.1 Key subject

The key subject is namely the internal subject of the school, including all the teaching and administrative staff and students. They are the active subject living and working in the school and they know and love the school the most. They are the group with the highest commitment level for this school in the society. They are the builders and reformers of the school culture and the internal impetus and the source of the school development. Their pivotal role displays in two aspects. First, they are designers of school culture scheme, as if without their participation and identification, even a good scheme is also useless. Second, they are implementers of school culture scheme and if they don't act or implement it, the scheme is also useless.

4.2.1.2 Secondary subject

The secondary subject refers to the relevant stakeholders of school development, including audiences outside the school like parents of students, personnel around the adjacent community, administrative department of education, expert team of university, schoolfellow etc. No matter how deep of their degree of participation is and how powerful their guiding force is, they are also the secondary subject for the formulation of the school culture scheme in the synergic sense of the school development as they belong to the external cause of school development. What assistances these secondary subjects provide for school development are: university expert teams repeatedly listen to the school demand, communicate with the school again and again and formulate the scheme of school culture development. Although they are byliners and planners, what they express and write are still the school's wills. Their thoughts and ideas can only be put into the scheme with the approval of the school. Administrative department of education is ususally the sponsor, urger, supervisor and evaluator of the medium and long-term development plan of the school and it can't replace the school to make a decision. The administrative department of education is the legal administrative subject and school is the legal subject of school. The former cannot impose any unreasonable administrative wills to the latter. Parents of the students and the community residents contribute their intellectual resources through participating in the formulation of school culture scheme and offer social resources and platforms all they can for the student activities at the same time. Schoolfellows are the unique wealth of the school; they can provide the school with social capitals like resources in politics, economy and culture. And they can also extend and increase the radius and energy of the school's social capitals in social relations.

4.2.2 Formulation Principles

What visual angle and height should we consider, establish and formulate a scheme for school culture construction in practice? Based on the years of related experience, we think it should follow the following three principles.

4.2.2.1 Connection and link

The principle of connection and link refers to that the formulation of school culture scheme shall be connected both horizontally and vertically. Horizontal connection and link means to unite the cultural keynotes of teacher culture, student culture, management team culture, parents' culture and community culture into a whole. Vertical connection and link is to throughout consider the history state, the current state and the future orientation of the school and to throughout consider the macro environment like the present social ideology and the international cultural background together with the regional culture and the micro environment of the school. The lantern model (as shown in Figure 4.1) put forward by Wu Zhongping can well express the connection and link principle.[3]

---

[3]Wu Zhongping Conflict and Mergence: New Perspective of School Culture Construction [M]. Shanghai: Shanghai Sanlian Bookstore, 2006: 12–18.

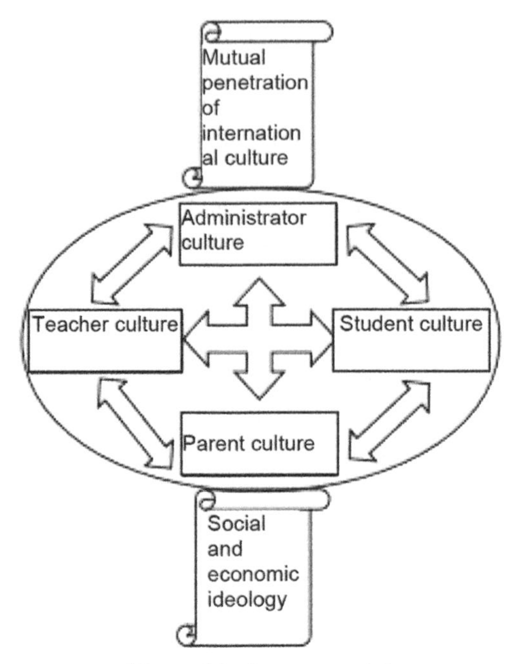

Figure 4.1 Lantern model.

The lantern model divides the principal influencing factors of school culture conflicts into school internal factors and school external factors. The four different group cultures inside the school are respectively teacher culture, student culture, parent culture and administrator culture. The principal influencing factors outside the school are the mutual penetration of eastern and western culture and the influence of the social and economic ideology. We have made two modifications to the lantern model (as shown in Figure 4.2), one is that we have added community culture into the original lantern model to show community is a key group that will affect the development of the school culture; the other is that we place the five groups inside the school side by side to show the relationship among the five groups. The relationship is that to take the student culture as the core, teacher culture and administrator culture as the school internal culture and the parent culture and the community culture as the extension of the school culture.

Based on the analysis of the influencing factors of school culture conflict, we should take a full consideration of these influencing factors and their mutual relations so as to be connected and well-informed while formulating school culture scheme.

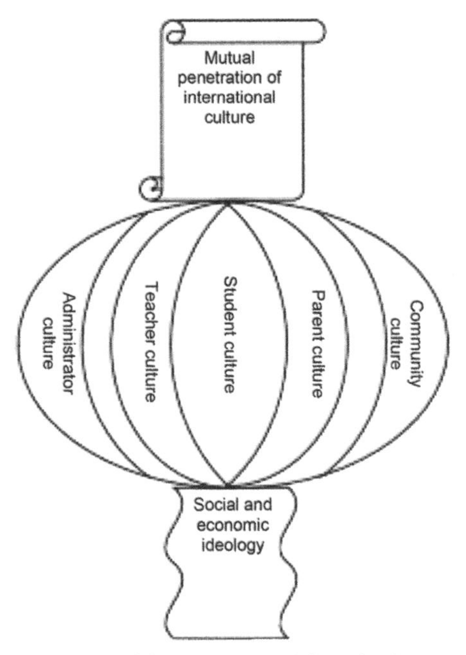

Figure 4.2 Lantern model (revised).

4.2.2.2 Multiple coordination

Multiple coordination refers to that the formulating process of school culture scheme shall involve multiple subjects and balance the opinions and suggestions of all the stakeholders, so as to draw on the wisdom of the masses and pool the wisdom and efforts of everyone. Take the revised lantern model as the basis, the scheme planning shall put students' development as the core and set the benefits of students' healthy growth and teachers' professional development as the top priority. For this purpose, we must handle the relationship between parents and communities properly so as to obtain a good reputation and support. The culture scheme shall be formulated based on the requirements of national education policies and the administrative department of education. University expert team is the third party without high-relevant benefits or high-risk relationship with the school. Their suggestions are direct and efficient, rational and exalted and thorough and complete. It is the important strength for the formulation of school culture scheme.

The second meaning of the multiple coordination principle refers to that the formulation of school culture scheme shall abide by the principle of inter-coordination between man and object, event and logic. Diversity and richness are encouraged by culture, but it must be the coordinated multiples extended based on the same logical system to avoid management troubles resulted from disorders.

4.2.2.3 System and completeness

System principle refers to that the establishment process of school culture scheme is a process that the university experts and the headmasters have a systematic learning together. The driver model of the school culture tries to bring the headmaster tools of methodology and epistemology of systematic school culture development rather than the knowledge and operation techniques of the school culture. If they have mastered the methodology of systematic school culture development, everything the school does will be cultural, otherwise, it only does things. Completeness principle refers to that the content of school culture scheme must be the complete two parts or four parts, namely the concept system of running the school (spiritual culture of the school) and practice system of running the school (include system culture, behavior culture and material culture) should be logic, ordered and expressed in the language style.

4.2.3 Formulation Process

The formulation process of school culture scheme is to sort out, consider, discuss and design a concept system and a practice system of running a school, which contains the participation of multiple subjects. It also needs the school to try its best, to make use of social capitals and to strengthen the solidarity of all members. For the school development, this is crucial, the making process can be divided into 10 parts and 4 steps. The 4 steps are completing the scheme, transforming the scheme into a plan, discussing in groups and passing the plan in the school meeting.

4.2.3.1 Completing the scheme

The accomplishment of the school culture scheme is one of the intervention strategies and variables of the school culture driving model as well as one of the formations of the achievement of the cooperative program. In the model of trilateral cooperation, 3 steps are related to the accomplishment of the scheme.

Step 1: intervention of the experts. After the administrative department for education successfully introduces the university teams, the member of the team will be distributed to different schools of the program. At least one expert, one graduate assistant and in some schools one teaching researcher will form a support team to help the school work. The task of the team is very clear: evaluating the school culture development with the principal team. And the leaders and managers of the administrative department for education will directly take part in this activity.

Step 2: organizing a team. The school which participates in the program should organize a team of school culture construction with the president of the school as the leader. And the number of the team should be decided according to the conditions of the school. Besides the members decided by the school, the members in the team of experts should join the team, too. The team of school culture construction is the core strength for making the scheme and its major responsibility is writing and discussing the plan, collecting relevant information.

Step 3: finishing the first draft. In our program, it usually takes us 6 months to finish the first draft of the scheme. And it is a significant, systematic and a complete scheme written by the experts of universities, fed back to presidents and schools repeatedly and discussed by the team of school culture construction. But it is more advocated and encouraged that the president and the school should do the job independently without experts'help. The intervention of the experts is not indispensable.

4.2.3.2 Transformation into a plan

Step 4: transforming the scheme into a plan is turning the scheme into a development plan of the school. The administrative department for education has strict requirements for the school in making a medium and long term development plan, which conforms to the medium and long term development plan of China and culture-strengthening policy. It is advised that the school transforms the school culture scheme into a school development plan during the process of making the development plan about the theme of school culture. The transformation is very simple. First, transferring the format according to the requirements with a clear purpose; second, turning the practice system of culture plan into a specific action plan with deadlines, persons in charge and forms of deliverable, etc. Meanwhile, it is necessary to make some specific and subordinate plans under the general plan, such as plans about the system culture, behavior culture and the material culture.

4.2.3.3 Discussing in groups

The first draft, presented by the team of school culture construction, both the first draft of school culture plan and the first draft of development planning about the theme of the school culture, are just in embryo and needed to be discussed by the members of the school and other related interest groups. Usually, the discussion will last for several months to half a year.

Step 5: discussing in teachers. The members of the team of school culture construction separately organize a group of teachers to discuss the plan and carefully collect information. It can be a group of teachers from the same subject or from the same course. Each teacher must give their opinions and suggestions, especially on the curriculum culture, teaching culture, researching culture and the construction of the classroom culture and student culture.

Step 6: discussing in students. The head teacher organizes the students for them to discuss and give their opinions on the ceremony of the school, student activities and other parts which are related to students' study and life. The team of school culture construction should collect all quantitative and qualitative data.

Step 7: discussing in parents and communities. The school should invite students' parents and the representatives of the communities to participate in the forum of the school development planning and the culture planning, with words and sounds recorded.

Step 8: revising the plan. After the collection of materials mentioned above, the team of school culture construction should revise and complement the school culture plan and the development plan. It will be done over and over again: asking for opinions, revising them, asking for opinions again and revising again. The final draft will not be formed until all members of the school are satisfied.

4.2.3.4 Passing the plan in the school meeting

Step 9: when the final draft is decided, it will be declared and passed in the faculty delegates' congress and become an outline, guideline, and action plan for the development of the school. The next step is putting it into effect.

Step 10: adjusting the plan. The above steps are procedures and steps of making the school culture plan. When it is put into practice, it needs to be further revised, adjusted and refined based on the practice information. Thus, making a school culture plan is a spiral and iterative process, which is the crystallization of collective wisdom. The whole process of making a school culture plan is shown in Figure 4.3.[4]

---

[4]Yu Qingchen: Culturology of School [M]. Beijing: Beijing Normal University Press, 2010: 180.

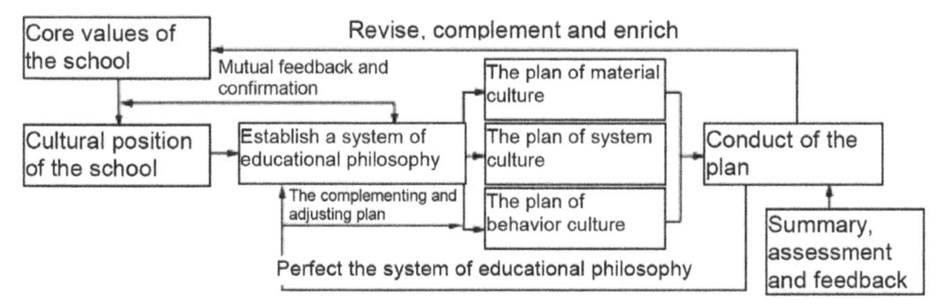

Figure 4.3 The process of making a school culture.

4.3 Contents of School Culture Scheme

Contents of school culture scheme mean what it tells and how it does. An integrated school culture program includes the planning purpose, the proposal of school's cultural icons or concepts, the explanation of school culture concept, school core value system, school practical teaching system, the related instructions, appendix and references. There are two points needed to explain.

First, when the school culture scheme is finished, it can be presented in two forms: a simple one and an integrated one. The simple draft is clearly framed and easy to use while the integrated one is attached with more specific plans which can be more easily implemented.

Second, when the school culture program is finished, it can not only be transformed into a school development plan but also be refined into a draft for headmaster's propaganda or for publishing in the newspaper. The title of the draft may contain a headline and a subtitle. Usually, the headline may be school core value, school motto, or the statement or slogan of school training goals and the subtitle is something about the explanation of school culture and its related practice and thought. Four parts are refined and remained in the propaganda speech or the publishing draft: proposal of school culture brand, explanation of school culture brand, school core value system (also educational philosophy system) and school practical teaching system. In the planning program, the purpose and process and other related information are also needed to explain.

4.3.1 Planning Purpose and Principles

At the beginning of school culture scheme, it is required to introduce the planning purpose and explain the goal of developing school culture to pave the following introductions. In the current educational circle, school culture construction has been an active demand for many schools' development. Nowadays, the school hardware are standardized and the school operating conditions are relatively in a balance, so the school competitions are more the competitions of education quality, school-running features and school-running quality. If a school wants to be the school with high quality, it must make a long-term plan for its school culture.

How to integrate the socialist core value into the practice of education? How to implement the strategy of invigorating China through culture? Under the influence of mainstream ideology, every school must have its own specific program purpose. For example, a new school of high starting point can focus on school cultural management directly; a school that has insisted on one certain educational view for many years needs to comb and simplify its school culture system; and if a school comes across the bottleneck in its development, the only way is to turn to the culture revolution or start with changing interpersonal culture or task culture. One more case is that if a school is not satisfied with status quo and willing to make changes, it should start with coagulating its core value or reengineering the process. No matter which case it is, they all belong to the scope of school development and improvement, turning into the direct motivation for the program of developing culture.

Every school has its own distinctive conditions so that their scheme bases are also different. Once the school has a clear planning purpose, it needs to formulate the feasible planning ideas or principles. Those principles must show a full respect for the cultural practice and cultural tradition presented in the process of school development. For example, the planning ideas and principles of Beijing Dinghuili Primary School's culture program which is called Dinghui Education program are as follows:

Respect the facts. Planning is not groundless. To respect the facts means that we must respect the historical tradition, the empirical facts and the related concepts of school culture development during the process of planning. The facts include the behavior of naming our school as Dinghuili and making explanation for it; the experience of carrying out the happy reading project; the design of campus landscape and the proposal and practice of Dinghui stars etc. Our school planning program will retain the original core ideas, but the utterance and expression will be changed.

Focus on the core. It means that when we comb the numerous slogans of school, we need to find out the core value of our school culture. Now, the school slogans vary and we lack the core value. Maybe the reason is that the core value is submerged by too many unimportant things. They are ordered illogically.

To be logical. It means that the core value is decided first, and based on it as the axis and core, the core value system is designed and added which includes the training goals, educational goals, school motto and signs etc. All of then must make sure that their logical relationships are consistent with each other in both content and forms. Many slogans that do not belong to the core value system must be abandoned. For example, we can handle the original explanation of the three characters, "Ding", "Hui", "li", like this: the meaning of the words "steadfast" and "wise" can be reflected in "calm the mind and behave intelligently"; and it is improper to replace "reasonable" with "polite", so the character "li" should be the original meaning.

Reconcile intension and interest. It means that we need to handle the relationship with the calm of "Ding" and "Hui" and the liveness of school and children. After all, education is not the Buddhism, school is not the temple and children there are lively, so there must be a feeling of heavy sensation if we require them with "Ding" and "Hui". How to reconcile the relationship between calm and liveness? Our program takes advantage of "happiness" to reach the balance on the one hand, we try to calm the mind and behave intelligently; on the other hand, we also need to live a happy life. The beauty is that the word "happiness" has also reflected the feature of our school's "happy reading project".

4.3.2 Proposal of School Culture

4.3.2.1 Concept of school culture

The concept of school culture refers to an abstract summarization of school culture.

In other words, the meaning is to name it, to find a cultural label for it, and to position its characters of school culture and structural features, which is just like someone's name that could reflect his personalities.

4.3.2.2 Condensation of the concept of school culture

Label of school culture is a combination of highly condensed school culture practice and proposals. In the process of condensation, four measures are commonly adopted. The first is considering the meaning behind school names, such as "Cuiwei Education" of Beijing Cuiwei Primary School, "Peixing Education" of Peixing Primary School, "Dinghui Education" of Dinghuili Primary School, and "Gengyun Education" of Gengyun Experimental School.

The second comes from school's typical projects, which has to deal with the problem of transforming features of typical projects into that of school culture, namely, the thought and practice of interests of Liberal arts is the foundation of Liberal arts Education of Hefei Tunxi Road Primary School, Harmonious Education comes from the Go project which has been dedicated to more than ten years by Beijing Fengtai Eighth Primary School, Azure Education comes from Beijing Haidian District Eastern Primary School's outstanding school-based school course of azure action. The third originates from school's selected proposals and routine practice that has formed a system for a long time. For instance, the 26-year-old Happy Education of Primary School of BeijingNo.1 Normal School can be directly used now. Hefei No.46 middle school always advocates the thought of integrity, so Integrity Education becomes its first choice. Primary school of Capital Normal College has always advocated the education that cultivates and develops children with child-like innocence. As a result, Child-like Innocence Education is naturally extracted. The fourth originates from regional impacts. Just like the flower culture of Beijing Fengtai District has been affected by the flower market around it. You'an Men No.1 Primary School consolidated two schools, the eastern one and the western one, putting forward the harmony culture and build a school environment that reflects the combination of Chinese and Western style. But it's needed to be pointed out that cultural label is not essential, if a school takes it as an unnecessary part, it doesn't need to condense it.

4.3.2.3 Proposal of school culture

This part will describe the evidence of establishing culture label from theory to practice. All the evidence should be solid, believable and comprehensive, which generally includes the historical source of allusions, school's tradition, school's teaching experience and proposals, school's source of students, the literature review of a culture label, the frequency of utilization of the culture label and specific material describing it, opinions of teachers, and university expert's proposals and suggestions.

Apparently, this part aims to describe the process of positioning the quality of school culture and explain the source of inspiration. Generally, establishing school culture label is not an independent task, and it is usually a respect, extraction and construction of school's inherent culture. The establishment of school culture label can't be separated from both its inherent tradition and the active participation of school culture plan's key and secondary subjects.

In May, 2012, we presented the idea and plan of Azure Education to Beijing Haidian District Eastern primary school. In the following, we talk about the reason why we came up with the idea of Azure Education.

Azure education is such an education that cultivates students with the spirit of confidence, tolerance, endeavor and freedom. Eastern Primary School's school culture can be summarized and labeled by "Azure Education". The reasons are as follows:

In terms of its source of students, the majority are children of migrant workers in cities, 83% of them with a non-local residence registration. Students come from different region and social class, leading to their huge difference in learning habit, learning foundation and parenting styles. The introduction of "Azure Education" attaches importance to the tolerance of the ocean, which means respecting every student's personality, being tolerant of their differences, and providing education for all students without discrimination.

The school culture label should correspond to ocean education and course features. "Ocean" is the label of Eastern Primary School, so taking "Azure Education" as guidance conforms to the developmental idea of the school. Since Li Ning became the principal of Eastern Primary School, she has made unremitting efforts to upgrade and improve the school. She is dedicated to creating a school atmosphere that embodies "ocean spirit", making every student feel a sense of being respected, every teacher a sense of democracy, and everyone a sense of group. The introduction of "Azure Education" is a reflection of the Eastern Primary School's spirit, that is, ocean spirit, namely confidence, tolerance, endeavor and freedom, which happens to coincide with principal's intention to extract and deepen ocean course. Eastern Primary School has widely applied the "ocean" to teaching, school-based courses, construction of course system and student activities. In the teaching field, the school has formed distinctive "ocean course" and combined it with other courses, establishing an overall outlook and focusing on the improvement of student's comprehensive quality. This includes the integration of various subjects, of subjects and activities, and of different content within the same subject. The school has written and published its school-based textbook: *approach the ocean know the ocean explore the ocean, and protect the ocean.* The school has built up an ocean detachment and a Young Pioneers group, and carried out "Red Scarf Volunteers in Action", telling stories of ocean to visitors in marine museum, popularizing marine knowledge, propagandizing protection of marine animals, advocating consciousness of environmental protection and low-carbon life. In this way, visitor's self-consciousness of "caring marine animals and protecting our earth" is enhanced. At the same time, volunteers give out leaflet, hand-copied papers and paintings in nearby community, kindergarten and geracomium, and preach Beijing Spirit and civilized life through display panel. Volunteers put the Spirit of Lei Feng into practice with their enthusiasm, making the capital take a pride of them. The introduction of "Azure Education" integrates its existing practice to some extent, establishing a system with inner relation between institution and activities.

The results achieved by experts' condensation and agreement. In March and April, 2012, Professor Zhang Dongjiao came to Eastern Primary School three times successively to participate in the discussion of construction of school culture. She reached a consensus with the school that it should turn features of marine projects into its characteristic of school culture, using the azure image to convey the spirit of ocean. In 26th April, 2012, expert panel of guidance from Beijing Normal University came to Eastern School to carry out workday activity. The experts present reached a unanimous agreement on "Azure Education", and put forward further annotation and suggestion of it.

4.3.3 Explanation to the Concept of School Culture

Explanation to the concept of school culture actually not only interprets connotations of school culture label, and responds to questions of "a kind of education" or "a kind of culture", but also plays a role of main points. In this sense, a system of theory on school management and a practice system are operation variables of this concept.

A brief and powerful sentence shall be adopted to interpret and express connotations of school culture concept with clear goals. From practical experience of running school, the goals usually include several possibilities, such as regarding cultivation of people as a goal, or core values of the school as core of narration, or regarding core values, training objectives and running school objectives as a goal, or two of these. For example, the goal of "Cuiwei Education" is to cultivate sunshine boys with illustrious virtue and ability of practice. "Tongxin Education" is a kind of education which regards sincerity, care and pursuit as the core values, which cultivates students with virtues of sincerity and care, which is to keep childlike innocence of the students through lots of advice from teachers.

At the same time, reliable sentences which are accordance with allusions, traditional disciplines and practical experience shall be adopted to interpret and express connotations of school culture concept. For example, "Dinghui Education" of Beijing Dinghuili Elementary School is put forward on the basis of:

4.3.3.1 Historical and traditional accumulations of the school

Dinghui Elementary School is a young school with much space to develop and improve itself. In August, 2008, the school took part in the project research of "characteristic construction of schools" which was launched by Haidian Education Commission to study again the topic of characteristic construction under the guidance of experts. In the process of summarizing, adjusting, perfecting and improving, a deeper insight and understanding to the characteristic construction of school gradually became clear. In this process, the school determines "Happy reading" project and regards "Cultivating students with happiness" as characteristics of the school. The school also determines the reading slogan, which is "Happy reading, promising future". The school fully explores connotations of the school name to extend the meaning of Dinghuili with three words including steadfastness, intelligence and politeness. The virtue of steadfastness means that students shall persistently devote all efforts to researches. The virtue of intelligence means that students shall have creative ability and continuously pursue more knowledge. And the virtue of politeness means that students shall be polite with others and hold an open-minded view to everything. And at the same time, the school also puts forward pursuits of teachers and students on the basis of these virtues.

At present, the reading project has given initial success eventually. On the basis of new development stage of the school and development history of school culture presented by leading group and teachers, the school tries to further improve the characteristics of this project to be overall characteristics of the school culture and also tries to explore connotations of the Dinghui education to form a concept of Dinghui education and a practical system.

4.3.3.2 Intellectual support of expert team

In April, 2012, a school culture project has been launched, which is created through collaboration of Dinghuili Elementary School and Zhang Dongjiao, a professor of Beijing Normal University. On April 20, 2012, workday activities has been carried out successfully, including school visit, presentation made by the headmaster, observation to classes, interview, and brain storm and so on. Many experts put forward a lot of constructive advice and suggestions to the process of brain storm. In the end, experts all agree that Dinghuili Elementary School chooses "Dinghui education" as the label of school culture.

4.3.3.3 Origins and enlightenments of "Ding" and "Hui"

Dinghuili Elementary School is named after Dinghui Temple. And the school name has an indissoluble bound with words of Buddhism. The meaning of Dinghui can be found in the Buddhism and classical expositions.

Two sources are found in relevant records by expert group. The point where to rest being known, the object of pursuit is then determined; and that being determined, a calm unperturbedness may be attained to. To that calmness there will succeed a tranquil repose. In that repose there may be careful deliberation, and that deliberation will be followed by the attainment of the desired end. These are seven levels of cultivation and certification in *Great Learning*. Buddhism was brought into China in Eastern Han. The Buddhism, Hinayana cultivates people into Arhat while Mahayana into Boddhisattva, and it also refers to these seven levels of cultivation and certification in *Great Learning*. The most basic principles of Hinayana in Buddhism are ethical conduct, concentration, and wisdom. It also combines calm unperturbedness and tranquil repose to form Buddhism term called Samadhi or Sampati in former translation, which means combination between calm unperturbedness and intelligence and also is basically equal to total meaning of attainment, the object of pursuit, tranquil repose, calmness, root, careful deliberation and desired end in *Great Learning*. The process from attainment, the object of pursuit to tranquil repose, calmness, and root is the level of calmness, while careful deliberation and desired end are well-spread results of Hui Guan. Mortality can be realized by cultivation with great concentration and then ereignis in *Great Learning* can really be close to.

In the principles of ethical conduct, concentration, and wisdom of Buddhism, ethical conduct is behavior conducts and rules for disciples of Buddhism; concentration means persistence and careful deliberation for Buddhism, while wisdom shall be applied into comprehending the truth of the universe. Threefold division is concluded from the noble eightfold path, which is the essence of practice Buddhism and the necessary step for disciples to get liberation.

Ethical conduct is the first step of threefold division which is made by Buddhism for these monk disciples or not. The kinds of ethical conduct consist of five precepts, eight precepts, ten precepts, upasampada, Boddhisattva precepts and so on. If a disciple followed the ethical conduct, he will physically and mentally be full of calmness and good virtue without any distracting thoughts and then he will reach to a state of selflessness.

Concentration is the second step of threefold division which means persistence and careful deliberation for Buddhism. Ethical conduct is guide sign made by Buddhism for these disciples. However, functions of the guide sign to change natures of disciples will be realized only by these disciples' concentration, or the functions are nothing. At the same time, surmounting the length of life in Buddhism is nirvana that requires disciples should make restless state of the power of the will be calm down. Concentration mainly means persistence for Buddhism with the goal of reaching a state with pure spirit through careful deliberation. The meaning of concentration is paying all attention to something without any distraction. Concentration in Hinayana mainly includes dhyanamaula and four kinds of formless realm, while reading Buddhist texts and sunyata are mainly in Mahayana. When keeping still and in deep mediation, disciples could wipe out much annoyance of the life and gradually have a deeper mediation.

Wisdom is the top of threefold division which used to explore the truth of the universe. Following the ethical conduct, keeping still and deep mediation and becoming a Buddhist are on the basis of Buddhism wisdom which is used for exploring the truth of the universe and the mentalities of the society. So, Buddhism wisdom is rational basis and result of following the ethical conduct, keeping still and deep mediation.

Ethnical conduct, concentration and wisdom are concluded from the noble eightfold path which includes right view, right intention, right speech, right action, right livelihood, right effort, right mindfulness and right concentration. Physical behaviors are especially restrained by right speech, right action, right livelihood, and right effort in ethnical conduct. Mental status is strictly restrained by right concentration. Right guidance to understanding of man is given by right view, right intention and right mindfulness in wisdom.

We can gain following enlightenments from explanations of Dinghui in *Great Learning* and Buddhism. Concentration is for disciples' thought and will while wisdom is behavior wisdom including the mentalities of the society and understanding of life. Combined understanding of modern meaning with school meaning, selected key words consists of mediation, encouragement, mentality and understanding. So, "concentration and wisdom, a happy life" is regarded as core values of the school.

It's obvious to us that each word in school concept is highly refined and has special meaning. Core values system and practice system shall be combined in the process of explaining and understanding the meaning. For example,

Beijing Cuiwei Elementary School explains "illustrious virtue and ability of practice" into following meaning.[5]

"Illustrious virtue" is from "What the *Great Learning* teaches is: to illustrate illustrious virtue; to renovate the people; and to rest in the highest excellence" in *Great Learning*. The aim of *Great Learning* is to carry forward fair character, to reject the old for the new and to make everyone reach the most perfect state. "Cui" mainly has two meanings which are green of hope and precious jadeite which is the most precious jade. So, we give "Cui" the meaning of sincere soul and illustrious virtue.

"Practice earnestly" is from "A gentleman should study extensively, inquire prudently, think carefully, distinguish clearly, and practice earnestly" in *Doctrine of the Mean* and other Chinese philosophical thoughts including learning to meet practical needs, every little counts, nipping in the blossom. These thoughts all reflects practical and innovative spirit.

4.3.4 School Core Value System

4.3.4.1 Interpretation of school core value system

School core value system is also known as school educational philosophy system or spiritual and cultural system, including core elements such as school core value, educational aim, school-running goal, school motto and logo, which are indispensable from the school. On this basis we can add some important elements, such as school spirit and school song. See Chapter 1 for details.

4.3.4.2 Planning essentials of school core value system

The first is systems thinking ability, which is the core of the whole plan. Core value is located in the highest logical position. Planners shall have the ability to refine and decompose concepts and the ability to understand the concept of orderly use and to expand logic.

The second is principles and laws of expression. There are several ways can be chosen. One is combining new way with old way and expressing according to the planning logic. The other is distinguishing new way from old way and starting thinking according to this logic: the presentation of the school's original proposal, the proposal we suggested and its interpretation, and the reasons why we choose this proposal or why this proposal is right.

[5]Zhang Yanxiang. Virtue leads to perfection, and working perseveringly in micro: Practice and Thinking from Cultural Concept of Cuiwei Education[C]. Statements of headmasters forum in Haidian District on December 11, 2014.

For example, when we were planning the core value of school philosophy of school-running Dinghui Education, we expressed as follows:

Status quo: there are various slogans in school, and one keeps pace with another, without core value.

Suggestion: express the core value from two aspects—pacifying mind and aspiration, emphasizing behavior and wisdom so as to enjoy life.

Interpretation: "Dingxin"—stabilizing your mind and aspiration. "Huixing"—acquire behavior and wisdom: be sensible and understand life. Pacifying your mind and aspiration and emphasizing on behavior and wisdom. The person who is pacifying his mind and aspiration and emphasizing his behavior and wisdom has the ability to enjoy life.

Reasons: it is rhythmic and readable. It is easy to read, spread and improve the popularity and reputation of school. It conforms to the new curriculum standard which focuses on cultivating emotion, attitude and value of students and the essence and spirit of Dinghui. It also conforms to causal logic—pacifying mind and aspiration and emphasizing behavior and wisdom, so as to enjoy life.

The third is mind, effort and point in place, which is aimed specially at school logo. Our team is a cultural team of university, not a culture company. What our responsibility is mind, effort and point in place: we help school to decide, whether the current logo conforms to core value, what are the improper elements, how to adjust, how to reinterpret to express core value on the premise that we don't need to redesign. If redesign is required, school needs to design it itself based on the suggestions mentioned above or from the school members, or entrust it to a culture company.

4.3.5 School-Running Practice System

4.3.5.1 Interpretation of school-running practice system

In accordance with structural theory of logic tree and cultural tripod of school culture system, school culture system is composed of spirit culture, system culture, behavior culture and substance culture. System culture, behavior culture and substance culture constitute school-running practice system. During the practical thought and plan, we integrate system culture, behavior culture and substance culture, constituting several important areas of school work: management, curriculum, classroom teaching, teachers' development, student growth and activity (including moral education), and campus physical environment. We can add or substract some areas or divide new areas in accordance with school specific conditions. See Chapter 1 for details.

4.3.5.2 Planning essentials of school-running practice system

The first is planning one by one. We expand it one by one according to several chosen areas. Every work area is the position to implement core value system, which can be called planning point.

The second is principles and laws of expression. There are various expressions of planning, without a fixed model. We suggest expanding every planning point through this logic—it is clear mind, and is easy to be accepted by school, such as the presentation of the school's original or current experience, the proposal we suggested and its interpretation, and the reasons and basis why we choose this proposal. For example, we suggest that the development of management culture and class culture of Xiangdong Primary School, Haidian District, Beijing Municipality, should be flexible management and self-confidence class.

Flexible management conforms to the characteristic of oceanic spirit and the actual situation of Xiangdong Primary School. There are three layers of meaning. The first is the integration of ocean knowledge and various subjects, namely, the knowledge integration and subject understanding. The second is the integration of science and poetry, namely, the flexibility of scientific spirit and humanistic spirit. The third is the flexibility of management.

Self-confidence or appreciation class is aimed at the fact that teachers and students lack self-confidence, focusing on the transformation of self-confidence as subject term into characteristic of class culture. Or it also can develop self-confidence from the perspective of appreciation, called appreciation class which is completely consistent with the spirit of Azure Education. School can discuss and choose by itself. We suggest respectively establishing class behavior standards for teachers and students. Answering and implementing the types of classes relies on self-confidence class.

Beijing FangCaoDi International School presents a proposal of harmonious and prosperous education. In May 2012, the project team made the following proposals for teachers' development of this school.

This school proposes three versions, including famous teacher, enlightened teacher and expert teacher. It is a realm and a pursuit. We suggest abolishing the version of expert teacher or clarifying the relationship between the expert teacher and the famous teacher and enlightened teacher. The famous teacher and enlightened teacher can be reserved. This is the meaning of expert teacher of FangCaoDi International School. And it also has answered the question, namely, what is expert teacher? This version is new and novel, signifying that the school possesses a harmonious environment and strong teaching faculties. Every teacher should be enlightened, and willing to understand themselves, others and the world. When it is the time, an enlightened teacher will have their name made. Enlightened teacher thus can be used in the management lists of teachers' honor.

Fangguyuan Primary School, Fengtai District, Beijing Municipality, advocates exquisite teaching and class. The project team makes the following proposals for its precision management.

Suggestion I: system process can add the form of graphs to make it clearer and more visual. We can divide the whole teaching and management work of school into several procedures, integrate the original texts before classifying them into relevant steps, distinguish the responsible person and competent department of various works in this procedure, and form teaching management system.

Suggestion II: it should form a special discursive style of exquisite education in system discourse. When expressing, it should be precise and fine, beautiful and vivid. We can put our efforts on system discourse. Exquisite education should have its own quotations and discursive style. We also should concentrate more effort on beauty to create beautiful campus environment, people and discourse.

4.3.5.3 Proper detail and clear responsibility

The planning process is a process in which university staff converse, discuss and consult with middle and primary school staff. The middle and primary school will ask for a more detailed plan and often consult how to implement this plan. At this point, we can illustrate with a pyramid (see Figure 4.4).

Figure 4.4 Planning pyramid of school-running practice system.

The top of this pyramid is school core value system, which should be completed by the planner. The planner can be from a middle or primary school, or from a university or somewhere else. The key of this part is the thinking ability of planner. The middle part is school-running practice system. It is mainly completed and written by the planner, but it also needs the help of the school, to consult with each other. What does this part test is the communication ability of the planner with the school. In this sense, it can be said that, the task can only be completed with the cooperation of the above part. The lowest part of this pyramid is action plan. On the basis of completing planning program, school needs to implement it by transforming it into specific action plan or executive plant. The key of this part is executive ability of school. There is a popular phrase in research area of executive ability: I prefer to adopt a third-rate plan and a first-rate executive ability rather than accept a first-rate plan and a third-rate executive ability. Of course, all organizers expect the combination of the first-rate plan and the first-rate executive ability most. During the transformation and execution of a plan, school still can communicate, consult and negotiate with the planner.

In addition, an integrated scheme should also comprise illustrations of relevant items.

The following is a complete school culture scheme.

Liberal arts spirit, children's world

—The thinking and scheming of theory on school management and practice system, liberal arts education, Tunxi Road Primary School, Hefei City.

Tunxi Road Primary School in Hefei, a school with three districts now, was set up in 1958. There totally are 3,756 students and 198 teachers (including 27 of them hired by the school itself) in the school. The covering area of Tunxi Road campus is 16,905 square meters, and the overall building area is 10,100 square meters, on which there are 37 classes, 1,869 students and 103 teachers now. The covering area of Century Sunshine Garden campus is about 10,000 square meters, on which there are 20 classes, 909 students and 45 teachers now. The covering area of Binhu campus is about 20,000 square meters, on which there are 20 classes, 978 students and 50 teachers now. On March 27, 2012, the project team of Beijing Normal University carried out the cultural construction working day in Tunxi Road Primary School. The initial impression for the school is:

• It's a growing school

This saying is given from the appearance of the school. The construction culture of the school has experienced three development stages: first, it's a school without anything, namely it's simple or unprofessional at the beginning. Second, it's a school with something, such as special places and classrooms. Third, it's a growing school. On the basis of being professional, buildings and classrooms are brighter, corridors are wide and it's also full of the humanistic care. Once coming in, you can find that Tunxi Road Primary School is a school especially built up for children. The font design is full of childishness and colors, and the wonderful bamboo forest square is vigorous. The children have enough room for development and also are confident and brave. It's a school, vigorous and active.

In recent ten years, this school has been developing fast and has been co-built by several parties to become a well-known school with good reputation. The school volunteer system is mature and parents are willing to participate in.

• It's a busy school

The school usually accepts inspections, and has various reception tasks. It is the Window School in Anhui Province. The pace here is very quick. All teachers and students are very busy. The slogan and idea of the school express the human-oriented spirit. However, the school actually focuses on matters instead of human. In the interview, all female teachers say something interesting that they all don't wear high-heeled shoes anymore, because they walk quickly without high-heeled shoes. Of course, only excessive workload can win you great honor and good reputation.

• It's a school with cultural self-consciousness

Fei Xiaotong puts forward the concept of cultural self-consciousness, including cultural self-knowing, self-evidence and self-leading culture. The school is clear in teasing its own cultural development stage and characteristics, and rich in the school-running philosophy system, e.g. the specific formulation aim of developing learned and refined Chinese in Liberal Arts is very striking. Cultural practice of the school is abundant and mature, and various projects there are of rich and colorful characteristics. The environment culture and buildings of the school are well-structured and well-arranged with a sense of design. The VI system is complete and distinct with appropriate colors.

While establishing school culture project in Baohe district, the school systematically synthesizes historical traditions and practical experience, based on opinions from experts of Beijing Normal University, leaders of the Education Bureau, teachers of the school and others, to choose liberal arts education to generalize cultural characteristics and practices of the school.

In June 2012, the experts group finished the first draft of school culture construction plan of Tunxi Road Primary School. On June 20, a seminar on planning scheme of the first draft of liberal arts education in Tunxi Road Primary School was held. On June 21, experts group made a third-time modification on the basis of absorbing the seminar spirit and rational suggestions, and then sent this draft to the school to supplement, modify and discuss. Half month later, a further modification was made and the final draft agreed by both parties was formed on July 31. On Aug. 6, leaders of the school went to Beijing Normal University again to have a discussion. On Aug. 23, they went to Tunxi Road Primary School and had an interpretation of school culture plan for the first time and carried out a discussion among all teachers. At the end of September, the final draft that both parties were satisfied with was formed.

4.3.6 Putting Forward of Liberal Arts Education

In recent years, the school gradually determines the value pursuit of "temperament and interest of liberal arts" and continuously deepens it through school culture construction practice on the basis of synthesizing the development of school culture and history. At present, this idea has been widely accepted by teachers, students and parents.

With an aim of cultivating national spirit of students, the school holds the classical reading activity with the topic of "Reading eternal essays, being modern teenagers". Develop children's humanistic qualities by reading the classics, and good behavioral habits by reciting traditional virtue mottoes, telling traditional virtue stories and joining in practical moral activities. Meanwhile, integrate reading ancient classical literary works and learning traditional culture and art together to encourage children to appreciate Chinese painting, calligraphy works, and classical melodies and learn Chinese chess to strengthen aesthetic education The school studies and select the contents among traditional culture that best fit children to learn, and compiles distinctive textbooks. By creating good school cultural atmosphere, students can enjoy it with freedom. All students and teachers can learn a lot of knowledge and learn from others' advantages. Erudition and elegance is a common goal of students and teachers of **Tunxi Road Primary School**.

At the same time, the school makes all members experience the importance of naivety by the activity of "Games Day". Meanwhile, educational idea that observes education and children from children's perspective is determined gradually. After establishing the "parent-teacher association", the school regularly holds parent-child activities in which all teachers, parents and children would participate, which makes everyone experience the importance of "love" in education. Ideas of "interest" and "love" are also rooted in the mind of everyone in Tunxi Road Primary School by various activities.

In March 2012, the project of school culture construction cooperated by Baohe Education Sports Bureau in Hefei and Zhang Dongjiao, a professor in Beijing Normal University started, with Tunxi Road Primary School being one of the project schools. On March 28, 2012, the working day activity of Tunxi Road Primary School was carried out smoothly. In the link of brain storm of working day, those experts gave much constructive advice and suggestions. At last, the experts reached an agreement that Tunxi Road Primary School choses "liberal arts education" as its label of school culture

In ancient Greek, Chinese characters "Boya" means "adaptive free man" in Latin. The so-called free man refers to those elites in social and political fields in ancient times. Ancient Greek advocates liberal education, with the goal of developing sense and cultivating soul, to help the students get rid of vulgarity and pursue excellence, but not to cultivate students to be tools through specialized training. Its education aims to free people but not slaves. With rapid development of modern industrial society, people pay more attention to the practical value of science and technology, which leads to a phenomenon in which people prefer to enhance the profit function of science and technology and the occurrence of deviation in education. Since 1960s, in order to highlight the mainstream value in western society, education system in Europe and America refocuses on humanistic education and takes liberal education as the representative.

Hirst from Cambridge University has profound influence on the definition of liberal arts education. He thinks that knowledge is a complex way to understand past experiences of people. The soul is not an entity that can develop automatically just depending on the inner rules of development, but the result formed after an individual absorbing human experiences that have become objective. The so-called soul construction process is exactly the process of absorbing various conceptual systems, and then, by taking advantage of experienced gained, understanding, introspecting and updating the world where the individual is in. Its core idea is as follows: first, knowledge is systematic achievements of human experiences. Experiences will be produced when human have connection with internal and external world. And these experiences can be expressed through language, namely the public symbols, and also can be inspected and corrected to be accurate. Second, the soul is not an entity that can develop automatically. The expression of soul activity is actually a result of knowledge, because it's brought in to the public by virtue of public symbols. So humans can perfect their soul and correct soul activity through knowledge. Third, knowledge is a result that human systematizes their experience through public symbols, Experiences have different forms after being classified. And these different forms should be found out and they are key points of liberal arts education.

The Logic of Education, written by Peters and Hirst, lists: (1) Standard for judging true or false and the deducible property of theorem system in the field of formal logic and math; (2) Natural science. Its judgment standard lies in observation of sense organs. (3) The awareness and comprehension of psychological states about oneself and others. In this field, unique concepts are trust, determination, attempts, wish and so on. (4) Moral knowledge and experience. (5) Aesthetic knowledge and experience. (6) Religious knowledge and experience. (7) Philosophy. Hirst emphasizes dimensional function of philosophy, namely the content of philosophical investigation based on knowledge and experience of other types. These various kinds of knowledge form course connotations of liberal arts education.

Thus, the classical meaning of liberal arts education is knowledge imparting, personal independence, and spiritual freedom.

We retrieve schools that adopt liberal arts education as the cultural label at present in domestic. These materials are mainly from the Internet, with only schools that issue school-running philosophy on the Internet included. So it is an incomplete statistics. On the basis of comprehensive selection, seven schools are screened at last, whose case materials about liberal arts education are obtained. These schools all put forward clear practice system and educational idea of liberal arts education, which have a certain influence and high reference value.

The interpretation of liberal arts education, in Yuqiao Primary School in Tongzhou District of Beijing is: "Liberal" is the structural construction of humans' love, emotional intelligence and noema, and knowledge ability, which pays attention to setting up a spirit of universal love, developing broad interests, effective guiding of knowledge and shaping various abilities. "Arts" is mainly for the construction of internal and external environment of the school, with the core of creating an atmosphere that is beneficial for cultivating teachers and students' taste and helpful for them studying in happy mood, so as to develop elegant temperament and refined taste in students. Liberal arts education of Yuqiao Primary School is to pursue "broad love, elegant emotion; broad knowledge, elegant behavior; various abilities, elegant atmosphere". Its goal is to develop teenagers with "sunny mentality and elegant temperament". On the basis of implementing quality education, promote students get comprehensive and health development, and shape sustainable development talents that have good character, elegant manner, refined taste, sunny mentality, balanced intelligence, universal interests, philanthropic heart and international vision. "Love" is a starting point of "Liberal arts education" which regards developing broad interests as the axis and treats creating a pleasant learning atmosphere as the core. Students can achieve success in mind intelligence and emotional intelligence by effective guidance to achieve knowledge, and the inspiration from combining emotional intelligence with mind intelligence. The integration of "liberal" and "arts" inherits and promotes the quality education, which not only embody characteristics of humanization, but also determine the direction in which school education gets all-round and characteristic development.

The school-running philosophy of Jusheng Primary School in Shunde District, Foshan City, Guangdong Province, is "Cultivating with liberal arts education, perfecting in morality and learning". "Liberal" is understood as broad thinking and richness, and requests students to develop broad interests and discover their own talents, which is a state to pursue knowledge; "arts" is understood as elegance and refining, and refers to elegant character and moral cultivation, which is a state of to pursue morality and interests. "Liberal arts" is not only to develop noble sentiments and healthy and rich spiritual world in students, but also make students have broad vision, broad interests and distinctive specialties.

Wenfeng Primary School in Wendeng City, Shandong Province, defines the basic meaning of literal arts education as four cores of "broad love, broad knowledge, elegance, and gentleness" and two levels of "knowing" and "action": liberal refers to profound knowledge firstly, including humanistic culture and scientific culture and innovation spirit and innovation ability. Arts refer to the philanthropic heart and noble sentiments internally, and are elegant temperament externally. Implementing liberal arts education is to combine being a person with being a successful person to develop a team of teachers and a group of students with profound knowledge, broad love, elegant temperament and elegant manners.

Wenhua Middle School in Wuchang City, Jiangxi Province, defines liberal arts education as: "it's a characteristic education, with the guidance of the Party and the country's education policy, goal of cultivating a new generation who are profound in knowledge and elegant in manners, good at cooperation, conscious in duty and responsibility, brave to innovate and also being able to face with all kinds of challenges in the 21st century; it's a kind of quality education that pursues profound knowledge and elegant temperament; the noble education in which the school if full of learning atmosphere, teachers are full of temperament of scholar and students are full of characteristics of elegance".

In addition, Caoyang No. 2 Middle School in Shanghai carries out liberal arts education with a topic of "arts and science share something in common, humanity being the leader". Nanjing No. 9 Middle School in Jiangsu Province takes "liberal arts and noble virtue" as its the essence of school culture. Deyang No. 5 Middle School in Sichuan Province regards "liberal" as the core of school idea, and regards "diverse education" and "liberal arts" as educational idea and school motto respectively.

At present, many schools adopt liberal arts education as their school-running characteristics, and also carry out lots of practice in aspects such as teacher resources construction, students' development, course installation, moral education activities and others through surrounding liberal arts. With the same liberal arts education, the interpretation of connotation with school culture characteristics is particularly important. Based on such interpretation, the practice system also has more profound basis and reliance. Combining with characteristics of the school, exploring and forming the school idea and practice system of "liberal arts education" with the characteristics of Tunxi Road Primary School are a key step in the process of school culture construction.

4.3.7 Planning Ideas

Tunxi Road Primary School is an excellent school with high social reputation and development platform. In the field of value, it pursues something profound and refined; in the field of cultural practice, something mature and abundant as well as unique are what it is looking for. The planning should be original, and great efforts should be taken to convince academically and logically. Under the premise of fully respecting, and using school culture facts and practice, the planning ideas and principles of Tunxi Road Primary School's Liberal Arts Education system are as follows.

4.3.7.1 Be concise and logic

As the ancient saying goes: Painting should be like a tree of autumn, which have not so many twigs, leaves, and it should also be like flowers in February which is different, unconventional or unorthodox, ahead of others. There are so many similar sayings related and their logics are parallel and independent. Hence it needs to establish a core value that is relatively stable and outstanding, on the basis of which, sorting the logic of core value system and making sure the forms, contents, forms and contents' logic unified and completed.

4.3.7.2 Combing the liberal arts education

The school has done very well in practice and accumulation of liberal arts education. Based on that, there are still two things to be done. The first is to turn the project characteristics of liberal arts education into the characteristics of school culture. The second is to clear up its source and meaning, make preparations for literature review and unify the logics of contents and expression.

4.3.7.3 Segmentation and unification

At present, the school has three forms of division in using the concept of liberal arts.

First, "Liberal Arts" is used separately in its meaning, and combined with "interest" simultaneously. Liberal arts interest has become a resounding slogan of school's characteristics. "Liberal Arts" means "learned and refined". "Learned and refined people are usually knowledgeable and polite." So long as we develop a good learning habit since childhood, acquire a wide range of knowledge and practice bravely, we can be knowledgeable. "Elegance" refers to "learned and refined". "Righteous person with virtue is elegant." Growing up in a profound traditional culture atmosphere since childhood, we may aim high and be elegant in the future. The connotation of "Interest" is "naivety". "A virtuous man is always as innocent as a child." Innocence and pureness as well as carefree heart are the most precious wealth of children. Let's engrave naivety on our mind and maintain an innocent heart forever. "Love" means "kindheartedness", "kindhearted people love others." When treating others like ourselves, we thus could become a moral person with the heart of love, great wisdom and personality charm.

Second, on the idea and practice, what draws people's attention are just students, but no teachers, which undoubtedly reveal that teachers and students are inadvertently isolated.

Third, the connotation of "Liberal Arts" gives its own interpretation, without using a profound meaning of the overall concept.

When planning liberal arts education, we shall obey the following principles. The pursuit and practice of liberal arts interest is the planning basis and should be preserved in ideas. Liberal arts education is a unity of teachers and students. Taking the traditional meaning of liberal arts education at the core, we shall elucidate its connotation and integrate its classical and modern connotations. Liberal arts education stresses scientific spirit while advocating humanistic spirit. Meanwhile, scientific spirit is also taken into account. So it needs to unify the humanistic spirit and scientific spirit, combining personal development with social responsibility, which is also liberal arts education should have.

At the same time, it also supports the idea of liberal arts Interest. Adopt the concept of liberal arts in fields, such as courses, moral education and teacher's management etc., and combine it with interest.

According to the textual research of the origin of liberal arts education and the conclusion of modern school practice, the meaning of liberal arts education established in this plan refers to the liberal arts spirit or liberal arts taste that is unified in humanistic spirit and scientific spirit. People who have liberal arts spirit or liberal arts taste is independent in personality and free in mind. Only such people can bear the social responsibility. By integrating its connotation, key words are concluded as follows: independence, truth seeking, equality, caritas, freedom and reasonability, which contain individual quality and social responsibility. People who have such quality are those with liberal arts spirit or liberal arts taste.

4.3.8 School-Running Philosophy System of Liberal Arts

Liberal arts education is the education that cultivates people with the quality of independence, equality, caritas, freedom and responsibility. The core value system of liberal arts education includes core values, training goals, educational goals, school motto, school logo and school stories etc.

The core values are the soul of school culture, which answer the problems that why school run education and values, beliefs, and assumptions behind it. With the core values being established, other formulations of value system will be solved easily. The core values and its expression must be the most superior in logic. It's not that the formulation of some core values is the idea of it. At present, there are three formulations in the school that are similar to the core values or at the logic position of the core values: first, cultivate learned and refined Chinese; second, the formulation of liberal arts interest; third, the idea that the school belongs to children and every child is important. These three formulations are parallel in logic. The third one just focuses on children, and is on suspicion of neglecting the teachers; While the first one separately use the meaning of Liberal Arts. It suggests using eight characters and two phrases to express the core values of liberal arts education: "Liberal Arts Spirit, Children's World". Liberal arts spirit is truth seeking, independence, freedom and responsibility. Children's World refers to a world that has not only rational brilliance, but also profound humanistic spirit, being full of children's interest.

Liberal arts spirit cares for the human's quality in training objectives, and children's world is corresponding to the school-running goals. In this expression, it not only covers the teachers and students, but also the connotation of "the school belongs to children, every child is important". "Children's World" is the simplification of the statement "the school belongs to children", which is catchy, easy to be read and memorized.

At present, the expression of school is that: cultivating learned and refined Chinese. It's resounding, and separately adopts the literal interpretation of two words of "Liberal Arts". Now, if we express the definition of training goals by taking "Liberal Arts" as an entirety, there are two kinds of expressions: One is in sentence form, which is cultivating the talented young with liberal arts spirit. The other is in the form of phrase, which is "Liberal Arts Spirit, Talented Young". This expression implies more profound meaning: Only the teacher with liberal arts spirit can cultivate the talented young with liberal arts spirit.

Presently, the school's school-running goal is expressed like this: Be people-oriented and construct the school to be the common spiritual homeland of teachers and students. The highlight of this expression is that the teachers and students are included in, and it can be closer to the meaning of liberal arts education. Liberal arts academy implies the rational spirit of truth and knowledge seeking, and personality spirit of equality, caritas and freedom. Children's world tells us this is a world full of fun under thenous. Liberal arts academy and children's world are also correlated with the metaphor of the spiritual home in original goals.

School motto is a common instruction for all teachers and students, and is also the most individualizing statement of school culture. Meanwhile, school motto is also the specific interpretation of the core values of school. There can be two choices of the school motto of Tunxi Road Primary School: The first is using the phrase "Liberal Arts". The second is using the statement that is completely consistent with the core values: Liberal Arts Spirit, Children's

World. The school could choose by itself or collect opinions from students, teachers, parents and famous people through organizing activities on school motto which accords with liberal arts spirit. The school should design all these by itself.

The school has no school song, so the information of school song, such as the name and the lyrics, should be completed.

School badge and its interpretation; the information that need to be provided by the school include the black and white pattern of school badge and the colored pattern; design specification of school badge; choice of basic color of school; present interpretation of school badge.

In order to express liberal arts spirit and the core values of Children's World, it is needed to have lively school stories as the carrier. What the school should do is to solicit stories from the whole school or parents, or the community. Ask the specialists to sort out two to three complete stories; the theme of stories are expressing liberal arts spirit (independence and truth seeking, equality and caritas, freedom and responsibility), Children's World, and Interest Young. The stories can be about teachers, students or both, as well as the historical development of the school, which are more convincing. It needs to deeply dig, and tell stories well. Stories should include those at present or in the past. It's recommended to take 10 years as a time period, and select a representative story every ten years. Every story represents a different theme.

Written by Diane E. Papalia and Sally W. Olds. Translated by Chenxi, et al, *The Children's World—from Infancy to Adolescence*, People's Education Press, published in 1984. It's suggested to find the English edition at the same time.

4.3.9 School-Running Practice System of Liberal Arts

The scheme of the school-running practice system of liberal arts consists of scheme suggestions on institutional culture, behavioral culture and physical environmental culture of liberal arts. The institutional culture, behavioral culture and material culture of liberal arts are the carriers of spiritual culture of the school. In order to correspond to the practice of the school, these three carriers of cultures are integrated and then divided into 6 fields: management, courses, teaching and classes, teachers' improvement, students' activities (including ceremonies) and campus environment. The school-running practice system of liberal arts of Tunxi Road Primary School consists of 6 parts, including consultation Mmnagement, courses of liberal arts, self-study classes, liberal arts teachers, sentiment and moral education, and Liberal Arts lyceum.

4.3.9.1 Consultation management

Negotiation management is the administrative management culture of Tunxi Road Primary School. The administrative management culture includes variables such as management philosophy, management system, and organization structure.

Management philosophy: the concepts of consultation management and communication management can be either adopted, and the school can choose one of them. Consultation management, or communication management, conforms to the liberal arts spirit of independence, equality, caritas and responsibility; also, it is set forth to direct at the current management pattern and problems in school. The negotiation between teachers and departments are not enough for the time being. The teachers complain about getting phone calls from different departments, sometimes getting several phone calls one day from one department. What worries the teachers most is the cancelation of the task they've just finishied.

Suggestions on administrative management culture contraposing these issues: the school and its two sub-campuses shall coordinate and communicate with each other so as to achieve the joint development; functional departments of each campus shall communicate with each other and integrate the functions and tasks so as to reduce burdens of the teachers; the functions of the internal teaching and research groups and group leaders of all grade groups shall be further improved, namely, the tasks can be transmitted to the group leaders instead of the teachers. The group leaders are in charge of the assignment of tasks to the teachers and the grade groups are responsible for the assignment and digestion of the tasks, which reinforces the administrative functions of the grade groups. To complete these improvements, the principal shall hold seminars and do researches, organize discussions among teachers to figure out the duties and work methods of every departments, and finally resolve the problems one by one.

Management system: by far, the management system of the school is developed sound and the *Management Manual of Tunxi Road Primary School* is formulated. The content of the manual is divided into seven sections, namely, teaching position, teaching, teaching research, class management, function classrooms management, fixed assets, and website management. Additionally, the regular administrative meeting system of Tunxi Road Primary School and the going-out learning system have been established in written form.

Suggestion: classify the systems; print the system texts and unified covers for them; the school logo shall be printed on the covers; evaluate the effect of the systems by reforming the procedures of one key system chosen. The principal is required to lead the teachers to participate in the improving process and practice.

Suggestions on organizational design: the principal shall assume overall responsibility for Tunxi Road Primary School. By far, Tunxi Road Primary School has three campus sites; the other two sub-campuses shall adopt vice-principal responsibility system; illustrate the overall organizational chart of the main part of the school; illustrate the organizational charts of the main campus and the sub-campuses; supplement materials, and illustrate all the existing functional departments.

4.3.9.2 The courses of liberal arts

By far, the special courses designed by the school and their descriptions are as follows (see Table 4.2).

The school-based course "Inheriting the splendid Chinese culture" has been on the list for 5 years.

It is ensured that the course will be available to every class once per week. During the course, Chinese teacher will carry out classic literature reciting and extracurricular reading. The Chinese teacher is required to draw up a reading plan before the commencement of the course, thereby making sure the course teaching is in order, activities are guaranteed, and examination and assessments are in place.

The Office of Education Research and Moral education Department shall cooperate with each other so as to make sure the regular implementation of relevant activities.

Sings of Tang poetry and appreciations of folk music shall be arranged in the music class every semester with proper teaching plan and class hour arrangement.

Traditional painting, calligraphy and ceramic craft shall be arranged in the art class every semester with proper teaching plan and class hour arrangement. The group leader of teaching and research shall make overall arrangement to make sure the intelligent use of special classrooms.

Table tennis shall be included in the P.E. class every semester with proper teaching plan and class hour arrangement. The group leader of teaching and research shall make overall arrangement to make sure the intelligent use of special classrooms.

Table 4.2 Special courses system of Tunxi Road Primary School

Subject	Content	Period	Teaching Requirements	Teaching Material	Assessment Method
Chinese	Calligraphy (Hard-pen Calligrapher and Brush calligraphy)	1 period per week; since grade three, Hard-pen Calligrapher and Brush calligraphy alternate every week	1. To help students cultivate proper writing postures and improve their writing qualities, thereby stimulating their enthusiasms in Chinese characters. 2. Students of low grade mainly practice calligraphy with pencils; students of middle and high grade start the practice with writing brush. All students have their special books. 3. Calligraphy Work Show shall be held by each class at least once per semester, during which the "Pioneer of Calligraphy" would be appraised and elected.	Special materials called *Xie Zi* published by Anhui Education Publishing House and the relevant content in the Chinese textbook	Student's calligraphy homework
Chinese	Recital	1 period per week, during which either	1. To stimulate student's interests in	*Enlightenment on Chinese Culture*	Reading plan of every class; design of

(Continued)

Table 4.2 Continued

Subject	Content	Period	Teaching Requirements	Teaching Material	Assessment Method
		classic reading or children's literature reading shall be arranged properly.	classic traditional Chinese literature and extracurricular reading and cultivate student's good reading habits. 2. To advocate no less than half hour's reading of students every day. 3. The course shall start with plans, supervision in the process of the course and assessment after the course is finished. 4. Every student keeps record of reading on reading sheet and the middle and high grade students shall take reading notes. 5. Every semester. The students and teachers are required to read at least one book jointly and hold a reading exchange meeting. At the end of the	published by Huangshan publishing house; *Read Eternal Articles and be Modern Youth* published by the school which are accessible in the library and reading corners of the classrooms and through other channel.	reading exchange meeting; students' reading record sheets

			semester, "Pioneer of book" would be appraised and elected.		
Mathematics	Approaching the mathe-matician	Mathematics class or class meeting	1. To cultivate student's interests in mathematics and help them get to know the mathematicians at home and abroad. 2. Mathematics activity class of "approaching the mathematicians" shall be held at least once per semester.	Optional materials	Teaching design
English	Pan Ying	7.5 periods for low grade students; 2 periods for middle and high grade students	1. To cultivate students' enthusiasm for English, learning English and using English. 2. The teaching shall comply with the special requirements of Pan Ying Teaching, especially the communication between school and parents and the assessment,	Beijing Normal University's special textbook	Second records of lesson preparation, teaching postscript, family study records sheet, monthly assessment record sheet

(Continued)

Table 4.2 Continued

Subject	Content	Period	Teaching Requirements	Teaching Material	Assessment Method
			also, the teaching for middle and high grade students shall be integrated with the average English teaching.		
Music	Sings of the Tang Poetry Appreciation of Classical Music	Music class	1. To inherit the traditional Chinese culture and cultivate students' classic temperament and foster their music literacy. 2. Make overall arrangements of teaching tasks for 12 semesters according to the existing materials with the characteristics of each learning phase. 3. At least one piece of Tang poem and music shall be taught per semester.	Optional materials	Students' performances and teaching arrangement and design
Fine Art	Ceramic Craft Traditional Chinese	Fine Art Class	1. To inherit the traditional Chinese culture and foster students'	Optional materials	Students' performances and works and teaching

	painting and its appreciation		capabilities of innovation and practice and improve their art literacy. 2. Make overall arrangements of teaching tasks for 12 semesters according to the existing materials with the characteristics of each learning phase. 3. At least one ceramic craft and painting thematic lesson which is no less than 2 class hours shall be finished and a ceramic craft works show and a traditional paintings show shall be held per semester.		arrangement and design
P.E.	Table Tennis	P.E. class	1. To cultivate students' consciousness of health and encourage them to take exercises	Optional materials	Student's performance table tennis competition teaching arrangement and design

(Continued)

Table 4.2 Continued

Subject	Content	Period	Teaching Requirements	Teaching Material	Assessment Method
			1 hour per day; let students master the basic skills of table tennis by systematic training. 2. No less than 4 periods of table tennis lessons shall be on the list and table tennis competition shall be held once per semester.		
Information	Computer	1 period per week (grade one and grade two)	1. To cultivate students' capabilities of using computer properly and safely and improve their skills of accessing information. 2. Set up folder for each student. Students shall accomplish their works at the end of the semester.	Optional materials	Student's performance and works, teaching design

Science	Three-Littles (little invention, little product, little paper)	Science class and interest-oriented class, or combined with the activity class of Young Pioneers	1. To cultivate students' capabilities of innovation and practice and stimulate their enthusiasm for science and the use of science. 2. To implement practical activities in compliance with the requirements of comprehensive practical activity. 3. To lead the students to finish at least one "Three Littles" activity and hold work show once	Consult the textbook *Comprehensive Practical Activity* and optional textbook	Students performance, Science and technology innovation works show
	Approaching the scientists		Implement at least once thematic activity per semester, forms unlimited.	Optional materials	Activity design
Moral Education	Chinese traditional motto story	Moral Education class	1. To cultivate students' virtues and inherit the fine Chinese traditions.	Optional materials	Students performance, teaching arrangement and design

(*Continued*)

Table 4.2 Continued

Subject	Content	Period	Teaching Requirements	Teaching Material	Assessment Method
			2. Make overall arrangements of teaching tasks for 12 semesters according to the characteristics of each learning phase by selecting the Chinese traditional stories scientifically. 3. No less than 2 motto stories per semester, and make sure the students can recite and tell them.		

Activities of "approaching the scientists" shall be included in the science class with proper teaching plan and class hour arrangement. Lectures shall be held on the basis of the trend of the times and scientific activities shall be implemented in order to improve student's' scientific literacies.

Every semester, all group leaders of teaching and research shall collect relevant materials, including the course scheme, teaching plans and teaching essays of all the school-based courses, and report them to the Office of Education Research.

With regarding to the above-mentioned factual circumstances and requirements of the core values of liberal arts education, I propose the following suggestions.

Liberal arts education shall fundamentally rely on its curriculum system. Thus the liberal arts curriculum system in Tunxi Road Primary School is designed and established based on the three-level school management system, namely national curriculum, local curriculum and school-based curriculum. Liberal arts curriculum system is the one capable of cultivating the spirits of liberal arts education in youngsters. This curriculum system aims at imparting valuable knowledge and nurturing core competence.

Richness. It means that liberal arts curriculum system shall be comprehensive, diversified and extensible. Under the development view of comprehensively cultivating humanistic quality and scientific rationality, liberal arts curriculum system designs diversified and extensive courses to accomplish school's training objectives.

Interestingness. It means that liberal arts curriculum system shall be interesting and full of interestingness. Liberal arts curriculum system shall be designed according to the interests and preferences of children and be interesting in both the content and the form.

Extensibility. It means that liberal arts curriculum system shall be extensible and inspiring. Liberal arts curriculum system is designed based on school's three-level curriculum management system, extending and integrating the contents of the three curriculums to the most within the limited time. It promotes the local and school-based curriculums with the national one. For example, calligraphy and reciting are, as extensions of the local and school-based curriculums, impressive in full swing.

Recessivity. It means that liberal arts curriculum system shall include some recessive courses. Liberal arts curriculum system is a generalized concept of curriculum, comprising explicit courses and recessive courses. The former includes subject curriculum and activity curriculum while the latter indicates those relational and cultural factors beyond the schedule but deeply influencing the growth of children.

How is the liberal arts curriculum system with the above four features like? If designing, processing and classifying based on the current situations of the school, liberal arts curriculum system shall comprise four curriculum modules, namely courses on extensive knowledge, courses on spirit and mind, courses on interests, and courses on social life. All the contents of the three-level courses are included in this classification and there are no absolute boundaries between any two of them.

Module I: courses on extensive knowledge. This curriculum module is to help students to acquire extensive knowledge and develop the spirit of truth seeking. It contains courses like Chinese, Mathematics, English, Science, Traditional Chinese Culture and Information Curriculum and their extended courses.

Module II: courses on spirit and mind. These courses aim at helping students to build up a strong body and a high spiritual taste and to free their minds. This module comprises courses like Music, PE, Fine Arts and Moral Education and their extended courses. School is suggested to extend the Moral Education to open etiquette course for students. Only students who are independent, confident and polite can meet the requirements of liberal arts education.

Module III: courses on interests. These are courses facilitating students discovering the interests of childhood and even life, embodying the idea that school is actually a world of children. This kind of courses is usually the same to the activity curriculum, also activities on interests and moral education in school.

Module IV: courses on social life. These are off-campus courses facilitating students observing social life and developing the consciousness and ability of being independent and responsible. This module includes courses of study tour, visiting and participation.

Suggestions on school-based curriculum construction: establish 1–3 complete school-based courses, including outline plan, teaching or reading materials, and evaluation on the courses etc. The school has begun to compile curriculum outline of classics reading and school-based teaching materials of etiquette course and chorus course in summer and the first draft is about to be completed in early September.

4.3.9.3 The classroom culture of self-education and self-study

The third supporting point of liberal arts education is classroom teaching, which is one of critical carriers for implementing and realizing the core value of liberal arts education. What kinds of classroom culture does the liberal arts education require? It's suggested to be designed from the following four aspects.

4.3.9.3.1 *First, name of the classroom culture—self-education and self-study*

The classroom culture of liberal arts education in Tunxi Road Primary School can be named as self-education and self-study. There are two reasons:

One is the accumulation of existing experience. In 2010, this school has been awarded as the educational research base in Hefei City. It is now conducting two major projects. One is *Study on Academic Evaluation of Pupils*, which emphases on the reform of final exam evaluation and covers all the teachers and the students. It includes sub-item tests for Chinese, layered tests for mathematics and attainment report of students. The other is *Self-education and Self-study*, which started in 2005 under the instruction of local education experts, Mr. He Bingzhang. This experiment that has the most participant teachers involves to following sub-projects: extracurricular reading, having nutrient breakfast-assuring sleep quality and insisting on everyday exercise, "self-study guiding" class, individual case analysis for students, being four masters and etc. At present, except the sub-project of "self-study guiding" class reformed is conducted in some experimental classes, other sub-projects are all conducted in every class. In the "self-study guiding" class, students have more time for self-study and teachers have more time for individual instructing and intellectualizing those student who need help. In such a class, every student join the process of autonomic studying rather than only teachers and a few students perform the teaching plan. The other is the thought of self-education and self-study meets the spirit of liberal arts which is beneficial to cultivate the awareness and ability of being realistic, independent and responsible for children, as well as challenges the awareness and ability of equality and caritas of teachers.

4.3.9.3.2 *Second, slogan of classroom culture—take responsibility from classroom*

This is the principal and value for teachers and students in liberal arts education to obey and pursue. Students seek the truth of knowledge through self-study, shoulder responsibility to solve problems independently or cooperatively, and experience the happiness of reason and freedom. Teachers should guide self-study, and challenge the comfortable field of existing experience to continuously explore and think how to treat children and solve problems in a higher point of view. Students are like the kites that can fly high and far while teachers, being like the string of the kite, must have responsibility and be responsible.

Third, evaluation standard for teachers in self-education and self-study class; it needs to supplement the existing practice in school and be further discussed or revised.

Fourth, evaluation standard for students in self-education and self-study class; it needs to supplement the existing practice in school and be further discussed or revised.

4.3.9.4 The liberal arts teacher

In order to encourage teachers to read, the school has developed *Reading Plan for Teachers in Tunxi Road Primary School* and equipped some public books for teachers such as *Advices to Teachers*, *Excellence Comes out of Teaching*, *Learn to Care*, *Love of Education*, *Humanities Reading for Teachers*, *19 Most Influential Pedagogy Works* and etc. Besides, books are allocated to teachers according to different disciplines. In 2001, the school set up educational theory study salon for young teachers in Tunxi Road primary school so as to lead teachers to read, create good reading environment and maintain teacher's initiatives in reading. Three years later, the word "young" has to be removed because more and more teachers joined in this organization. In the March of 2006, the school opened the blog of teachers in Tunxi Road Primary School; and it set up the BBS of Tunxi Road Primary School in December of 2008. In recent two years, the school further creates reading atmosphere by a series of measures such as building office bookshelf, subscribing books recommended by teachers initiatively, commending and rewarding "the office of books". Along with the introduction of systems, practicing of measures and establishment of platforms, teachers in Tunxi Road Primary School gather together to read, take notes, write stories as well as take part in discussions and teachers' BBS.

From the above data, we can draw the conclusion: the evaluation standards for teachers are detailed and quantified. Emphasizing reading is one of strategies for cultivating liberal arts teachers. There are many methods in the school. However, the problems that need to solve are: what is the relation with liberal arts teachers and how to avoid and reduce the hollowing out and impersonal risks caused by overusing systematic management.

Professional developments of teachers and team construction have always been the top priority task in the school, which is a long-term and hard work. Liberal arts education demands liberal arts teachers. With regard to the current situation of school, the behavioral culture development for liberal teachers can be considered from the following aspects.

First, set up standards for liberal arts teacher. School should firstly give the answer to the question that what kind of teacher is the liberal arts teacher. The standard of liberal arts teachers mentioned here mainly refers to the behavioral standard of teachers including quality and image standard and classroom behavior standard. Since classroom behavior standard has been discussed in the above section, we mainly discuss quality and image standard in this section. Quality standard: liberal arts spirit should be independent, realistic, equal, caritas, free and responsible. The person who embodies liberal arts spirit must be graceful, learned and refined. Therefore, the image standard for liberal arts teacher is: beautiful, regular, amiable and elegant in appearance. Uniforms for teachers and students are distinctive and beautiful, so it's suggested to hold uniform show on Teachers' Day. If possible, it could be considered to customize cheongsam for female teachers. They also can wear high heel shoes in proper situation.

Second, set up incentive system and honors management system for liberal arts teachers. The levels of incentives and honors can be divided into liberal arts teacher, erudite teacher, considerate teacher and humorous teacher. Among them, liberal arts teacher is the highest honor and the other three are sub-item titles. Evaluation and assessment standard and frequency can be set up individually to combine with administrative purposes, or confer honorary title instead of combining with administrative purposes. How to use the results should be determined through school discussion.

Third, cultivate and care about teachers differentially. Now the school has 198 teachers, including 1 national excellent teacher, 6 provincial excellent teachers, 22 municipal excellent teachers in Hefei and 16 district excellent teachers, which reflect considerable number and reliable quality. The school needs to seek higher development to cultivate 3 to 5 special-grade teachers. It's suggested to separately set up plans for cultivating special-grade teachers. Activities to care about teachers such as 30 years on the 3-feet platform activity or warm activity of my happy days in Tunxi Road Primary School should be designed to motivate the enthusiasm in senior teachers. Besides, hold teachers salon, and add and carry out activities related to profession development of teachers.

Fourth, start with investigation. Since this is a well-known school with high reputation, both the school and teachers here are busy. Teachers in interview reflect generally that they are so busy that they have no time to do professional work, which caused them to be fickle and hard to calm down or take a break. How to handle this problem? The answer is staring with investigation. The special guidance of time management is of tremendous assistance for improving work efficiency, which help teachers free from miscellaneous daily routines and reduce the job burnout of teachers. The school can establish an expert guidance team to make a school-wide investigation in the teaching affairs in one day or one week. Simplify and reconstruct teachers' daily routines, optimize teachers' time arrangement and improve teachers' ability in time management by applying the theory of time management and existing practical experience and coordinating with the consultation management work in the school. The problem behind current situation of teachers' time management may be the chaotic management, which need to be solved by deliberating and discussing by both the school and the teachers to clarify thinking, retrieve time, and get time back to teachers, so as to guide teachers to develop. Teachers need a rest.

Fifth, study on cultural development. By virtue of the method of process reengineering, ravel out the process and procedure of teaching and researching activity and lead group leaders and teachers to discuss how to organize the teaching and researching activity efficiently and how to improve it. Other problems such as the group cooperation in self-study experiment and effect evaluation can also be studied. It's suggested to adopt the method of brain storm and start from the least satisfied project.

Sixth, establish a new system for selecting the class teacher. At present, Chinese teachers are in charge of classes. New standard may be tried out: the position of class teacher can be assumed by those who is willing to be or the new teachers, but not abide by the standard based on discipline. Discipline based standard is against the liberal arts education spirit.

Seventh, use culture manual. The school has made a characteristic and elegant culture manual for liberal arts interest. After identifying and assuring the system of liberal arts education, a culture manual of liberal arts education should be prepared to disseminate the values to new teachers and students, which also can be one of efficient measures for foreign exchange and publicizing liberal arts education.

4.3.10 Liberal-Interest Moral Education

Student activities are one of the features and advantages of the school. It is of highly efficiency in two aspects. The first is a series of mature theme activities, including arts, festivals, science, games and so on, that the school has developed.

The traditional culture and art activities include reading classics (classic reading competition, Tang poetry sing competition, ancient poetry performance competition, etc.), reciting traditional virtue mottos, telling traditional virtue stories, taking part in practical moral activities such as Tunxi Road Primary School forum, classics discussing between teachers and students, class teacher tells traditional virtue stories, and protecting eggs activity, etc.; appreciating Chinese painting, calligraphy and classical instruments, for example, introduce folk music into school activities, etc.

The activities themed with traditional festivals include telling annual customs of traditional New Year, watching festival lanterns, guessing lantern riddles and rubbing Yuanxiao in Lantern Festival, and carrying out enjoying moon and guessing riddles activity in Mid-autumn Festival. All these activities not only arouse children's interest to Chinesc traditional festivals and enhance their awareness of inheriting of traditional cultures, but also expand children's horizon, which are beneficial to children's long-term development.

Series of activities in scientific innovation month include Science fiction painting competition, science composition competition, telling inventor's stories, visiting provincial and municipal science and technology museum, attending science lectures, technological innovation competition of "wonderful ideas" and little invention competition. This stimulates students' enthusiasm and interest in studying science.

"June 1 Game Day" is children's favorite day. The game Playright, which is introduced from Hong Kong, wins much popular among students. Each time after the game, students would create games by themselves and then make the new games popular among them, which helps them develop their ability in creation. On the one hand, the large scale activities expand teacher's horizon; on the other hand, it helps us to re-know children, pay attention to children's characters and needs, and deepen our understanding in children.

Rich and colorful activities in clubs such as Chinese chess, dough modeling and chanting verse give play to the functions of "two academies and two centers", namely Youth Science Academy, Youth Art Academy, Youth Art Activity Center and Youth Sports Center.

The second one is about resource utilization outside school. In 2007, Tunxi Road Primary School attempted to establish a new parents' organization—Parents and Teachers' Association. It is made up of teachers and parents. There are volunteer department, planning department, liaison department and others departments in it. All the positions are assumed by enthusiastic parents. The school leaders and teachers help parents to launch activities. Most of the plans and expenditure are solved by enthusiastic parents. Parents and Teachers' Association have planned many theme activities which greatly enrich the educational resources of the school. Parents get a better understanding in the school education and teachers through participating in school management and organizing activities. They are able to give sincere comments and suggestions, and actively help school to deal with difficulties. Parents establish reliable and harmonious relationship with the school and teachers. Such relationship integrates every child's family and the school education. It makes each child feel the care from both family and school.

The school sets up three kinds of course, including entire training, system training and selective course, which constitute the main learning content for parents and the school. In the entire training course, there are parents open week, parents' meeting, reading, eating, sleeping together with practicing promotion meeting and so on. The parents open week enables parents to know the class and know children's advantages and shortcomings. The parents' meeting is a platform for the parents and teachers to talk about the education. The annual experimental promotion meeting of "read every day, have nutrient breakfast, ensure sufficient sleep, and do exercise every day" aims at the parents of new pupils, namely grade one, to remind them to attach great importance to the shaping of good habits such as reading, eating, sleeping and exercising, so as to help the students to develop good habits. In the meeting, the psychologist, nutritionist and sports expert will give parents the scientific and systematic training.

In the systematic training of parents and school, the school thinks that family is an important base for reading promotion. However, the effect of arousing all parents to pay attention to reading cannot be achieved by only one or two report(s) or training(s). The school sets up parent-child reading corner. Each class recommends one family to take part in it. The school makes a set of training plan, which includes eight courses, covering the content of demonstration and guidance, theory studying, experience sharing, and experts face-to-face, etc. The course content focuses on children reading. After the first phase of training, those families, together with class teachers, will organize parent-child reading activity. A single spark can start a prairie fire.

In consideration of some parents have work in the day time, the training organized by school is held in the night. In the bright library, the parents and children sit close together to read under the guidance of teachers, forming the warmest scene in the night.

As for the selective course, the school moral education department would discuss course arrangement with the principal of parents' school and director of teaching and discipline each semester, guaranteeing one expert lecture, one theme for each lecture, each month. The lecture content covers reading promotion, child psychology, study habits and how to deal with children's rebellious behavior in adolescence, etc. The school informs parents the time and content of lecture through fetion and Home-School Communication. At the same time, prepare class permit at the gate keeper's place. Thus, parents who want to attend the lecture can take the class permit on their own will. The parents can decide whether to attend the lecture or not according to the content, and also can choose courses based on interest.

From the data of the five theme activities, we can conclude that the principal of Tunxi Road primary school is busy, the teachers are busy, and the students are busier. The busy life reveals that this school is rich in fund and is high in concern extent. The activities are self-contained and interesting. They are good for the liberation of the soul and the release of personality. Those advantages should be well maintained. And on this basis, we should coordinate these activities with the help of the concept of liberal-interest moral education. The suggestions are as follows:

Firstly, the explanation of liberal-interest moral education. Establish liberal-interest moral education system of characteristics of Tunxi Road Primary School based on the liberal arts spirit and core values of children's world. Morality is a kind wisdom in interpersonal communication. Liberal-interest moral education is an education form to cultivate children to be independent, fair, and kind, as well as help them to experience and percept the beautifulness in life, which belongs to the course on interest. Liberal-interest moral education aims at enjoying the happiness of life in the process of pursuing rational brilliance, and making the children physically and mentally healthy. Through activity curriculum and potential curriculum, we hope that children are capable of living lively.

Secondly, slogan of liberal-interest moral education. "Freedom comes from contact" can be taken as a slogan, which tells its relationship with liberal arts education and liberal arts spirit. This slogan can match with the slogan of class culture "Responsibility begins from class." It reads catchy and pleasant to hear.

Thirdly, autonomous control of class. It advocates carrying out student self-management and class autonomous control system. What the school should do is to make self-management and autonomous-control regulations. The system can be different from class to class. Each class can own different class motto and name. The regulation ensures every child has the opportunity to take responsibility and show himself, for instance, the selection and assumption of flag raisers, class leader and group leader in turn, as well as the management of books in library. This enables every child to get the opportunity to be the master of class. The class is not the world of excellent students but the cradle of every student. Meanwhile, we should pay attention to the growth of boys.

Fourthly, the establishment of behavior standard of model student. It is feasible to set up honorary title of liberal arts youth, which is in accordance with the honor and award system for teachers. The standard can be made according to each grade respectively. And then the honorary title is appraised and elected once a semester. The school should encourage teachers and students to discuss the standard and make it.

Fifthly, the design of personalized rituals and ceremonies. There are various colorful activities and daily ceremonies in school. It is necessary to design personalized rituals and ceremonies that can represent the liberal arts education. I have two pieces of advice for it. One is the establishment of "say hello everyday ceremony", the other is the performance of "Ode to liberal arts" in some important occasions which include flag-raising ceremony on Monday, orientation and graduation ceremony, important activities and so on. It requires all the students and teachers to recite together. It is better to wear school uniform when holding important activities. I write initial content of "Ode to liberal arts". (The school can encourage teachers, students or invite famous people to write it). For reference:

Freedom comes from contact, responsibility begins from class.

Children's world, liberal arts school.

Liberal arts courses, liberal-interest moral Education.

Be independent and realistic, be equal and philanthropic.

Be responsible, self-study and self-education.

Liberal arts spirit, talented teenager.

Sixthly, emphasis on key activities. There are many special activities in school. Some parents put forward that whether the theme of these activities can be highlighted, whether some activities can be normalized or serialized, and whether some different activities can be themed the same. And at the same time, some activities can be deleted appropriately after discussion. For Tunxi Road Primary School, it's necessary to discuss what the school needn't do.

In 2008, the school firstly employed specialized Culture Design Company as a counselor for brand image to design VI brand image in Hefei. The school hopes every wall, every corner and every square can be made use of to educate students.

The school goal of liberal arts education is "liberal arts school, children's world". The topic is in line with beautiful environment and well-equipped school where there are interesting bamboo forest square, talking story wall, golden dragon recording reading, reading desks with a gap and game room for lower-grade pupils. What the school needs to do is to name the school after liberal arts school and make some explanation.

4.3.10.1 Relevant matters

1. This is a detailed map about cultural management of Liberal arts liberal arts education system, which involves all work areas in the school. All related issues have corresponding plans and advice here. Based on this, the school can attach various sub-schemes texts on it. These sub-schemes are a part of executive process of school's cultural scheme, so a special person is needed to take charge of, adjust and complete the work.
2. This is a first draft that needs to be widely discussed and carefully researched by all members of the school. Even each element in the train of thought in planning, principles, and core values should be comprehended and discussed fully, so as to reach the consensus. This is just the process of cultural management and leading.
3. This is a planning scheme of school's cultural development. It focuses on helping the school untangle the logical thought of the core value system and sort out the cultural practice as well. On this basis, the school needs to work out an implementation scheme or an action plan so as to assign and carry out things that have reached a consensus.
4. Just one cultural scheme does not mean the cultural construction can be accomplished in an action. That's because good ideas need to be performed, and the act of performing can neither be replaced nor copied. Everyone in the school is a rational agent in cultural management and

construction. The planning scheme of liberal arts education will have to gain recognition and support from all the faculty members and students of the school; meanwhile, the implementation plan needs to formulate relevant supporting measures, and be constantly improved, adjusted and enriched according to new situations during the process. Make sure that managers and operating personnel get trained in the process of implementation to help the leadership, management cadres, teachers and students learn the connotation and significance of liberal arts education, and constantly improve the consciousness and initiative in implementing the scheme.

After some stages of the implementation, rational and rigorous assessment and feedback shall be provided in time so as to constantly improve the quality of work, and realize the goal of cultural development that the school has expected: Liberal Arts Spirit, Children's World.

5. Tunxi Road Primary School is a new school with high starting point and high expectation, expecting to continuously create brilliant achievements through liberal arts education. Guiding the school-running practice with systematic and completed core value system can shorten the course of struggle for several years, so as to develop better and more effectively. After gradually completing the intended target and plan, the school may continue to improve the construction of spiritual culture, material culture, system culture and behavior culture, forming a complete system of school culture based on the brand of liberal arts education.
6. This planning scheme can directly serve as the basic blueprint and basis for the formulation of school's five-year plan of development or school's development plan.
7. Liberal arts education system thoroughly conveys the characteristics of school culture, which further promotes the social popularity and reputation of the school. As a result, it greatly enhances the school's core competitiveness and gradually makes the school become one of the nation's elite schools. The overall planning scheme, serving as the cultural tool for school's improvement and development, can promote the management level and teaching level of leadership and management team as well as the team of teachers.
8. All of the advice in the planning scheme is only for the school's reference, but not to force it to accept. The right to choose and make decisions is entirely up to the school.
9. After the scheme is discussed and confirmed, it enters the implementation phase with the school as the main body. In fact, the process of soliciting

options and discussing already represents the beginning of the school culture construction. The principal needs to turn it into an all-people movement, not a one-man fighting. The formulation and implementation of each sub-scheme and the management of each key event must be participated in by people who are appropriate enough and of different levels.

10. Culture is a long way to go but also a. There will be many difficulties and risks in the construction process, therefore risk assessment must be conducted; it is neither right to assume that it will run smooth all the way nor blindly trust experts.
11. The panel of experts reserves the right to use the scheme text as teaching cases.

5

Advancement of School Culture Practice

In fact, the school has always been an existence of culture, since it has been engaged in conducting activities of culture practice. From the aspect of experiment and study, school culture-driven model forms a powerful intervention. Therefore, for the convenience in study, we set the start and end time of school improvement project as the boundary time for experimental study. In school culture-driven model and improvement model of trilateral-cooperation school, the process of advancement of school culture practice just starts from the formulation of school culture scheme, or the time when the project of school culture construction is launched. From a broad sense of school culture improvement, the advancement of school culture practices includes four aspects, which are setting goals of school culture construction, formulating and choosing school culture scheme, carrying out school culture scheme, and evaluating school culture and practice effect. According to the study and exposition aforementioned, it is known that the school culture and its schemes cover various aspects in school work, which is especially true when they are transformed into the action plan that guides the school development. Therefore, the advancement of school culture practice studied in this Chapter mainly refers to exercising, implementation, being practicable and evaluation after the school culture scheme is affirmed. And matters on formulating school culture scheme have been analyzed in Chapter 4. In this Chapter, the implementation process in advancement of school culture practice will be mainly discussed, which includes three links of the dissemination and identity of philosophy, determination and implementation of the scheme, evaluation and feedback of result, and answers will be respectively provided to three corresponding questions: What will we do? How will do? How about our performance? This Chapter will continue to develop the exposition, from the perspective of school culture improvement, on the advancement of school culture practice from two aspects of content and scheme.

5.1 Dissemination and Identity of Philosophy

What's the formation mechanism of school identity? From the relation between culture and identity, one of basic functions of culture is to provide the symbol system that the humans all accept. These symbols can be seen from beliefs, knowledge, languages, and history to external aspects such as behavioral habits, dressing, and hairstyles. For example, two persons from the same culture tradition or culture patterns share a common set of symbol system, with some affirmation, identity, appreciation or cordial feeling on the common culture in their mind. If being integrated, these mental feelings can be called as "identity" or "cultural identity", which is mainly to answer the question of "Who are we?" The culture provides a fundamental reference system for humans. In real social life, the vast existence of the deviation of "we/you" or "we/them" is just the result of cultural identity.

The relation of school culture and school identity is derived from the relation of culture and identity, which is the specific reflection and application of the theory of relation of culture and identity in the field of school life. For all students and staff in the school, school identity doesn't exist from the very beginning, but is cultivated and comes into being gradually in the process of school activities. However, what is noteworthy is that students and staff in the school not always generate school identity in their process of study and work. The generation of school identity is influenced by various factors, but has no direct relation with the time of staying in school. For example, some school leaders, teachers or students have no good identity on the school though they have stayed there for a long time. They are not only void of correct understanding on the school and the school culture, short of a kind of affirmation, reception, appreciation and even active blending, but also have some circumstances that some deflected cultural concept occurs in their behaviors. There is another phenomenon that some students and staff in the school have some school identity, but the identity is not that strong; or they may have no identity in the core value system of the school, but just have the identity in external part of the school culture, such as the school environment culture.

Core value system of a school is the soul of school culture, which should be reflected in every aspects of school work and be familiar to, understood and practiced by all students and staff. The most important aspect in school identity is the identity in school core ideology. In those days, Xiaozhuang Normal School founded by Tao Xingzhi was exemplary in cultivating such identity. If a school has very good core values and core value system, but it pays no attention to these and is slighting in dissemination, and just keeps these in the mind of school leader team and on school files, such core values and core value system cannot play the role in bringing people together or enhancing identity. Therefore, it's a necessary work that needs to be persisted in for a long term to keep cultivating, promoting and enhancing the identity of all students and staff, as well as the public, on the school through disseminating school cultural concept inside and outside.

The formation and identity of philosophy means disseminating and marketing school culture construction scheme, especially the part of school-running philosophy system, to the school public and social public in an active way by taking advantage of various methods to form core values that is steady, abided by all members of the school and accepted by the social public, so as to enhance the power of school culture and promote the school culture to move forward towards higher school culture spectrum. Make all students and staff of the school and the public know: what will we (the school) do and what's the expected effect?

5.1.1 Dissemination Content

For school culture planning scheme, the dissemination content is mainly the school culture concept and school core value system (namely the school spiritual culture system)—the explanation of school core values and school logo and so on. In fact, such dissemination work starts in the process when all staff members participate in the discussion of planning scheme. Researcher and compiler of the scheme, as well as managers of the school, all know such a principle: the lower the barycenter of decision is and the stronger the potency dimension is, the fewer barriers will be. However, in the early process of making culture scheme, the participation of the public is of fragment and dynamic nature, and the content is also of high uncertainty and variability. The purpose of dissemination is to make people form a complete impression on school philosophy system and practical operation structure, settle differences, acquire consistent identity, and improve cohesive force, popularity and reputation of the school, so as to form favorable public opinion and cultural image.

5.1.2 Identity Scheme

Identity refers to the converging process in emotion and mentality of individuals and others, groups or imitation characters. The identity of school culture philosophy refers to the process in which all students and staff of the school and the public gradually accept, believe in and finally voluntarily to bear the responsibility of school culture construction and then become willing to make their own efforts in the school spiritual culture system and practical operation structure from unknowing or even doubting in this. If the main emphasis of school culture scheme is set low, more people will participate in; the dissemination will be easier; identity degree will be higher; and the difficulty in implementation will be small. Four links and ten steps of making school culture scheme mentioned in Chapter 4 lay a good foundation for this. But, after the scheme is determined, philosophy dissemination scheme is needed to be made specially to look for identity strategy.

5.1.2.1 Make school culture manual

Dissemination via written media is one of most important dissemination ways, of which main forms include: school restricted publications, such as newsletter, teachers and students' newspapers and periodicals; wall newspaper, blackboard newspaper, glass-fronted billboard and etc.; information manual, such as culture propaganda manual, employee manual, welcome manual and training manual; suggestion box that the school leaders used to collect teachers and students' opinions, comments and suggestions; letters; slogans and announcements; questionnaires and etc.. Disseminate through school publications is an important information spreading means. The management principle of publicity materials is: know about the readers, be geared to the needs of the targeted public, add contents according to reading interests and psychological needs of the public; save print cost as much as possible; keep balanced layout; be particular about the design and properly illustrate with pictures.[1] Comparing with oral and audio-visual media, written

[1]Zhang Dongjiao, Public, affairs and image: Introductory Theory of School Public Relation Management [M]; Chongqing: Chongqing University Press, 2005: 116–117.

media is advantageous in the convenient in carrying and keeping publicity materials; besides, the targeted public are able to choose methods of intensive reading or quick viewing according to their own needs when reading the materials, which is beneficial for tehm to accurately understand the content and structure of school core value system. In this part, school culture manual is focused on.

School culture scheme is not only a guide to direction, but also a tool for dissemination and marketing. The scheme is able to be transformed into a school development plan themed with culture, the text of a speech for the headmaster, as well as being made into the school culture manual. While, its name can be school culture manual, school dissemination manual, or be named according to the concept of school culture or slogan. The content of this manual plays the role of dissemination and declaration for school-running philosophy system and school-running practice. In first part of the manual, contents such as the school core values, training objective, school-running objective, school motto, school song and school spirit must be displayed; while in its latter part, practical actions and achievements should be showed. It's necessary to make the culture manual delicate. It may not be thick, but it should be of good quality and distinctive characteristics. Select and use the qualified paper in making it. Indispensable elements on its cover include school logo, school name, the manual name and etc.. Besides, the type setting shall be reasonable. Adopt the format that matches pictures with characters; choose representative pictures elaborately and make sure no same pictures are used; try best to fully reflect the school culture philosophy and practical activities. For those schools that have entered into the continuous spectrum field of school culture, they always take school culture as one of curriculums. In such situation, in addition to culture manual, corresponding detailed school culture reading materials or teaching materials are needed to be developed. For example, Gao Huanxiang, headmaster of Jieyang Huaqiao High School, wrote school-based reading materials such as Campus Scenery and Calligraphy in School in person. Teachers and students there are able to blurt out the connotation of the school motto of "Ri Xin Qi De" (literally improve morality everyday), as well as tell the origin, purpose and source of the calligraphy used of the scenery "Sit under the hundred-year-old tree, read thousands of books". After being made, school culture manual can be reserved. Revise every two years. If it's the first time for a school to make culture manual, it'd better choose the period when the school culture scheme has been carried out for some time.

In addition for making school culture manual, school logo should be widely used in the office system and dissemination system in the school. Print and customize writing paper, envelopes, cups, desk calendars, scarves, time management manuals, handbags and etc.. on which the school logo or school motto is printed; besides, school logo can also be applied on the gate, hall door or classroom door, walls, school flags, PPT templates, school uniform, prizes and so on. Select three to five representative objects from these and put them into the handbag together with school culture manual to be ready to present to those who come to visit or are interested in the school, so as to make visitors remember and recall this school well.

5.1.2.2 Common identity among teachers and students

For internal dissemination of school-running philosophy, its objects are all students and staff, and its purpose is to promote the identity of school culture improvement system. School identity has a direct internal relation with school culture improvement and construction. On the one hand, for students and staff of the school, a strong school identity will arouse a common sense of belonging, pride and honor of "I belong to such-and-such school", and promote to form a high-level organizational commitment, which will evoke a kind of conscious, positive and high-degree sense of responsibility, so as to play a role of motive power to inspire them to work and study hard, as well as create a good atmosphere for school culture improvement and construction. On the other hand, a high sense of school identity will also enhance the sense of trust among all students and staff, reduce barriers and problems in exchanging and communication, and encourage all school members to actively participate in activities held in the school, so as to make the school culture construction the objective of struggle not only for school leader team, but also for all students and staff.

School culture manual and school culture scheme should be distributed to all staff in the school, with each of them holding one, so as to help them know well and blurt out core values of the school, school motto, school logo and its meaning, school-running objective and school spirit etc.. In the initial stage of dissemination of the newly organized philosophy system and practical operation structure, a quiz or competition would be helpful to or intensify the teachers' memory. This method seems to be unskillful, but it is of high efficiency; if some small gifts or bonus is provided, it will be more interesting. Such method is applied in one of our project schools. It turns out that teachers are interested in this and participate in in high spirit. The headmaster of this school once said something reasonable, "I know the testing method is stupid. But how could we take actions or carry out these relevant things without keeping them in mind? My principle is: remember and then do it."

Also, it's necessary to distribute school culture manual to students, with each of them holding one. Though this method increases the expense cost, it indeed deserves such investment. This is because the students would take the school dissemination manual home and share honors and achievements of the school together with their parents. For communicating school culture with students, teachers could do this through holding class meetings, having classes, guiding students to discuss their feelings, make suggestions and write compositions. In some schools, school culture would be made into school-based course. For example, in the school culture manual of Jieyang Huaqiao High School in Guangdong we mentioned above, all characters in campus scenery spots are facsimiled from the copybook written by famous calligraphers; besides, a school-based calligraphy textbook on the origin, features and calligraphers is composed. After new students and new teachers entering in the school, the first thing for them is to know about the school culture and requirements, which is the common way and standard of doing things for members there. Moreover, it's also practicable to lead the students to the school scenery spots where represent the school culture to explain and guide them to observe relevant content, as well as interpret cultural meaning of the bell or school applause. In some schools, all students and staff would shout slogans together, recite school motto or school goals together in the break, which is really imposing. Within half a month, each student and teacher would know about the core value system of the school and commit it to memory.

Developing education on school history for all students and staff is one of effective methods to conduct internal dissemination of school culture improvement system. Each school has its own development history which started from a historical time and is undergoing at present and will go into the future. School history is a kind of good materials for improving and enhancing the school identity of all students and staff. Carriers for manifesting school history are of various patterns. It's practicable to help those students and staff to continuously know about and review the school history by forms such as school commemoration day, homecoming day, and special issue of school anniversary, as well as express honor to those who made outstanding contributions to the school in history. What the most important thing in this is to abstract complete stories that is corresponding to the school core values and hand down these stories to make them become classics and legends. For example, Fengtai Fifth Primary School in Beijing takes the petrel spirit as the school spirit in its happiness education. It adopts such a story to explain its spirit: "About half a century ago, it was the very beginning when the school was just founded. Conditions there were poor and harsh. At that time, teachers took the school as their home, living there, and devoted themselves to work for it. Sometimes, they were so tired that they fell asleep without even taking shoes off." For another example, on the Writer's Wall in Beijing No. 2 Middle School, pictures of some well-known writers graduated from this school and their achievements are displayed there, which makes each visitor impressive and also realize "atmosphere cultivating people".

Disseminating through oral and audio-visual media is also a frequently-used method, of which main forms include conference, conversation, broadcast, video and etc.. Conference would gather people together and provide the chance to speak and listen at the same time, which is a two-way disseminative communication form, being open and controllable. An effective conference should have specific goals, complete and thorough plan and arrangement, and definite explanation; meanwhile, it also depends on the leadership and presentation skill[2]① of the conference chairman. The school is able to take advantage of various chances that are beneficial to disseminating school culture, such as the school staff meeting, school leaders' meeting, parents' meeting, conversazione and seminar, to show the importance and feasibility of school culture construction and great achievements for people. At the same time, the school can know about opinions and suggestions of the school staff and parents, which has a stimulative effect on the school culture improvement and construction. Conversation is a face-to-face communication method, which is very effective in exchanging ideas. Broadcast is an important and frequently-used method to inform important news and notices to all students and staff, which is also a good platform to disseminating school culture. Dynamic image materials such as documentary films can bring people vivid visual and auditory experience. Fine-made advertising video for school is equal to an exquisite visiting card for the school to display its spirituality.

In this part, let's focus on discussing the meeting forms that can be referred to in the school culture philosophy dissemination phase. First, interpretation meeting for school culture scheme. After the university contact and the school jointly study and complete the school culture scheme, the school is suggested to convene an interpretation meeting for it. In the meeting, the university contact is responsible for making the interpretation report. As the third party who combs and writes it, the university contact is able to objectively and comprehensively interpret the process of developing the scheme, proposal

[2]Zhang Dongjiao, Public, affairs and image: Introductory Theory of School Public Relation Management [M]; Chongqing: Chongqing University Press, 2005: 136.

of the school culture philosophy, formation and content of the school-running philosophy and its practice system, and so on. Second, press conference for the school-running philosophy system. Beijing Cuiwei Primary School is reported to have adopted such a meeting form. So, if there is enough faith and maturity in school culture, this form can be chosen. Third, taking advantage of the school anniversary. It is a great opportunity that should be seized, as all the public including parents and the government, communities and brother schools, experts, scholars and public officials, local public and international friends would all gather together in that occasion. The spokesman must be the principal. For example, Primary School Affiliated to Chinese Academy of Agricultural Sciences, Peixing Primary school and Fifth Primary School in Fengtai District of Beijing presented their philosophy of growth education, talents cultivation education and happiness education, as well as their practice systems in their school anniversaries, which achieved good effect. Fourth, competition for the school-running philosophy system. It can be held among students and teachers, respectively or collectively. Attractive award-wining questions, quizzes, questionnaires and games can be designed. Fifth, seminar for the efficacy. After the school culture scheme has been carried out for some time, and the above-mentioned forms have been tried, the school is suggested to conduct a survey on the school culture philosophy system and its practice system and collect data through ways of questionnaires and interview. After the analysis on such data, a seminar should be convened to comb and summarize the problems existing in the dissemination and implementation of the culture philosophy system, so as to make corresponding adjustments.

5.1.2.3 "First attack" on parents

When the school culture philosophy system becomes mature, through dissemination, the school can hear opinions and suggestions from people of all walks of life and expand its influence, which is advantageous for the school to gain more social resources to develop. In its dissemination, parents, communities and the public are the major targets.

Parents are the chief and primary public for the school, for they care most about the school's cultural environment and its influence on their children. Some parents who possess specialized knowledge in education care or worry very much about the school's description on its running purpose, educational objectives and school motto even observe its practice with expertise, and then would put forward opinions and suggestions. Students' parents are the first and primary target of the school cultural dissemination. There are four ways and strategies for the school to gain the approval of parents.

First, take advantage of the parents meeting. The parents meeting in each semester is a great opportunity for the school to disseminate its school-running philosophy, practice system and educational achievement. The school is suggested to try to seize such opportunity to insert the dissemination concisely and properly in the speech and activity arrangement. At the meeting, the spokesman should simplify the speech, in which the dissemination can be arranged at a proper time. In the first dissemination, each element of the school-running philosophy system and its supporting practice point should be systematically described.

Second, take advantage of the school culture manual. After the parents meeting, a handbag filled with school culture manual is suggested to be dispatched to each parent, so as to conduct simple training and disseminate by virtue of the culture manual.

Third, take advantage of forms of letter and school-parent communication. In the daily communication with students' parents, the school can adopt the forms of letter, message, and so on, and use envelops and PPT marked with school logo or motto to consciously express the school-running philosophy. What is the most important is that the teachers should always behave in accordance with the core values of the school in solving students' problems. Only in this way can its core values be remembered, identified and supported by the parents. Take Hefei No. 32 Middle School as an example: in the communication with students' parents, its teachers are able to exhibit specific examples to show their responsibility for students, which leads to the approval of its responsibility education by parents.

Fourth, take advantage of the parent committee. Each school has set up the parent committee at levels of school, grade and class, and some have established the parent volunteer system. They can be utilized in the following ways. First, in each term, the school should convene at least twice a periodic parents' meeting to solicit opinions and suggestions on school management and development from parents. Some parents who are experienced and have professional spirit are likely to provide constructive suggestions which are helpful in school development and management. Once, a sixth grader's father who held an important national post made a suggestion for the primary school: the educational goal of the school should be training students to be great Chinese, because he believed the world will belong to Chinese twenty years later. Second, the school can rationally make use of parents' enthusiasm and abilities by virtue of the volunteer system. Some parents with great enthusiasm and abilities are willing to contribute their wisdom, labor and money to the school, so the school needs to provide such a platform for them. The parent committee can make a research list to learn about parents' aspirations such as the time and project of the voluntary work, and the type of and energy for the service, as well as assign work for them: when they should attend reading class with the teacher or prepare reading materials, and when they ought to help the school receive inspection or visit. Take Fangguyuan Primary School as an example: on the working day of expert team in April, 2012, the experts were received by thoughtful parents who were wearing with uniform clothes and smile. At 5 pm, the principal asked them to go home, but they sticked to their post till the experts left. Third, the parents committee assists parents and the school in managing involved affairs. For example, it can hold lectures and regular meetings, help the school organize spring outing, and arrange another things.

5.1.2.4 Open up to the society

School is one of social organizations, so gaining acknowledgement from the society is certainly a pursuit of it. To open up to the society, there are three strategies and approaches that can be tried.

First, make use of school anniversaries and open-up days. School anniversary and its open-up days are those days on which the school has closer contact with publics outside such as the government, communities, parents, schoolfellows, international brother schools, etc. Plan the school anniversary and open-up day elaborately, so as to systematically expound the school-running philosophy system, as well as the cultural expression of visual and auditory identification identity system such as school logo, school flag, school song, story, ceremony, celebration, applause and school bell in a proper way in the main part of the celebration or performance. What needs to be noticed is that only when the school culture system is mature can it be demonstrated in the celebration of school anniversary systematically and ceremoniously. Otherwise, there will be a few opportunities to change, and something widely known can not easily be changed, because illogical statement can not only bring bad influence, but also cause public distrust to the school. This point mentioned above should also be noticed when marketing on the open-up day and on-site meeting. Open-up day and on-site meeting provide the school with an opportunity of self-presentation to the superior, counterparts and society. In such occasion, the self-presentation should be systematically in any case. Thus, systematic thinking is very important.

Second, make use of school network. As early as in 1998, Internet has been officially announced by the United Nations Information Committee as the fourth media emerged after traditional mass media like magazine, broadcast and television. With the rapid increasing of Internet users in our country, Internet at present has become an important tool for the school to disseminate its cultural concept, and more and more schools have built up their own website as a platform to disseminate their images to the outside. Through the construction and improvement of network platform, the school can update publically the construction and management progress of its school culture on time, and collect advice and suggestions for improvement from the outside world at the same time. Internet is also interactive. For example, discussion on Internet is substantially influential, because this kind of web text can not only be revised by the author, but also be removed and updated by the administrator, and can also be directly enclosed with added comments or revised result by readers, which promotes the initiative of participants and easily draw the public's attention.[3] Dissemination of school culture concept system or culture scheme can be conducted through various interactive ways of Internet, such as school forum, message board and blogs of teachers and students, to resolve problems existed in school culture concept system and strengthen publics' understanding of the school. Campus network can also be taken as a platform for communication between teachers and students. For some problems that are inconvenient to be put forward face to face, they can be discussed by means of the network, which can not only reduce conflicts but also enhance efficiency. Campus network is a direct tool to disseminate school culture. Thus, it is very important to manage network well, including designing reasonable layout, using school's basic color and updating information on time. On the home page of campus network, there must be declarations about the name of the school, name of each sector, school basic color, pictures, school's core values or school motto, which should be direct-viewing and catchy.

Third, make use of articles and media. The principal and teachers can publish research paper related to school culture, revised scheme of school culture or speech text of the principal on the educational newspapers and periodicals. By this way, more and more people will get to know about this school, and the popularity and reputation of the school will be highly promoted. To gain acknowledgement and support from the public, the school should not only strive to improve its education and teaching quality, but also take full advantage of the dissemination function of media. From the aspect of public relations of the school, dissemination needs to be encouraged, in which a principle "disseminate as long as something is done" should be abided by and

[3]Pan Hong. Theory and Practice on brand spread of primary and secondary school [M]. Chengdu: Sichuan Education Publishing House, 2007: 50.

the proportion of the dissemination and deed is 9:1. When the core concept system of the school becomes relatively mature, media dissemination (such as holding a press conference) can be made use of to elucidate explicitly to the public about the background, objective, and process of school culture improvement; meanwhile, pay attention to the feedback from the public. In a word, this is to make the school culture concept be widely acknowledged, win support from all walks of life to the uttermost, and popularize the school's brand.

Cuiwei Primary School in Haidian District of Beijing, since being selected in the school culture construction project in November 2008, has been positively using resources such as university experts, parents and communities, and finally built Cuiwei education brand and core value system that are mature and acknowledged by all teachers and students in 2010 after repeated discussions and argumentations. On April 2, 2010, Cuiwei Primary School convened a conference on core concept of school culture. At the conference, Zhang Yanxiang, principal of Cuiwei Primary School, elucidated explicitly about the background and process of determining the core concept of school culture, as well as the profound meaning contained in school-running objectives, educational objectives, school badge and school motto. Besides, he also answered questions about school's development from macroscopic perspective and microscopic perspective put forward by journalists. At the same time, parent representatives, student representatives and designers of school badge and school motto were invited to express their opinions respectively. Participants of this conference included leaders from Education Committee of Haidian District and Sub-district Office of Yangfangdian, friendly units in surroundings, parents, students, teacher representatives and over ten media representatives from CCTV, Beijing Youth Daily and Sina, etc.. The purpose of this conference was to "build new image, spread new concept, set new goal", lead the school to commence and run at a high level, and construct a widely acknowledged school brand of "Cuiwei education" during the new development period of Cuiwei Primary School, with innovative strategic vision, more ambitious objectives and more profound connotation. This was also the first formal press conference convened after the spokesman of education system in Haidian district took up his position, which attracted comprehensive concern from the society and media.[4]

---

[4]"Virtue leads to perfection, and working perseveringly in micro" Cuiwei Primary School spreads school culture. (2010-04-02) [2012-01-01]. http://www.china.com.cn/info/edu/2010-04/02/content-19739870.htm

5.2 Implementation and Execution of Scheme

As an important link of promoting the culture practice, the implementation and execution of the culture scheme refers to decomposing tasks to responsible persons and mobilizing the internal and external forces of the school to carry out each item of work of culture construction with the planning cultural scheme as model and basis. Even if the scheme is prepared perfectly, it will remain stagnant in the design phase forever without implementation and execution.

5.2.1 Contents to Implement

Implementing the scheme is actually to carry out the culture scheme and all its sub-schemes, including behavior culture construction scheme, system culture construction scheme, and material culture construction scheme or such sub-schemes that refer to work schemes in various areas as school management, teaching, curriculums, classes, teachers, students, and physical environment of the school. Such schemes can be decomposed more detailedly as required. For instance, the culture planning scheme of Chang'an Xincheng Primary School in Fengtai District of Beijing includes several sub-schemes such as the general planning scheme of "Wisdomsun" educational philosophy system and practice system, wisdom management action plan, "Sun & Wisdom" class culture action plan, sunshine campus construction plan (including the construction plan of the construction culture, design plan of the human landscape, management plan of school library and network), design plan of sunshine rites and ceremonies, action plan of moral education, and the design and choice action plan of school motto and badge. This is an example, the speech of the Principal, Chen Cuimin, of Beijing Fengtai District Donggaodi No. 3 Primary School at the culture promotion demonstration meeting on July 6, 2012.

5.2.1.1 Time and Space Rendering, Enculturation, and Endless Discovery

—Consideration on the culture construction and promotion of Donggaodi No. 3 Primary School.

Established in 1969, Donggaodi Third Primary School is located in Meiyuanli Community in Donggaodi. At present, it has 24 classes available, more than 800 students, and 55 teachers including 9 municipal and district-level backbone teachers. During its 43-year development history, it carries on the traditions and continues to innovate constantly. We know clearly that the culture construction can guide the development of the school, unite people, and encourage members involved to move ahead. Therefore, we constantly refine the cultural core, define the direction of development, and carry out the independent development way.

5.2.1.1.1 *Refining meticulously to form an educational system of "Time and Space Discovery Education"*

We carry out the spaceflight spirit education by making full use of the locational advantages of being adjacent to the China Academy of Launch Vehicle Technology, which achieved a good educational effect. On this basis, we make progress in inheritance, define educational ideas, and strive to construct the school into the cradle of future space-time explorers, taking the Time and Space Discovery Education as school-running philosophy, regarding cultivating the future space-time explorers who can create life value as educational objective and treating "extensive learning, pioneering, goodness and aesthetic education" as school motto. Our motto is that "space-time and discovery are both endless."

In 2008, we put forward the philosophy of "Time and Space Discovery Education", which is the best interpretation of "spaceflight characteristics", as well as the most accurate summary and promotion of the school culture.

We can understand the connotation of this educational philosophy from three levels as follows:

1. the discovery activity is the effective carrier of "Time and Space Discovery Education";
2. the discovery quality is the force origin of "Time and Space Discovery Education"; and
3. exploring the value of life is the pursuit of the essence of "Time and Space Discovery Education".

Therefore, it can be deemed that Time and Space Discovery Education is the education to cultivate the students' space-time discovery quality, inspire students to display vitality and guide students to create the value of life through the space-time discovery.

5.2.1.1.2 *Striving to fulfill to form the cultural characteristics of "Time and Space Discovery Education"*

Unique: be distinctive

The core philosophy of school curriculums based on "Time and Space Discovery Education" is unique: namely unique thinking method, unique creative inspiration, and unique displaying method.

Creative: great ideas

The students are able to carry out independent and distinctive creations when they have original ideas, in which the students can think hard and devote to creation.

Rich: Various curriculums

In order to achieve the objective of "Time and Space Discovery Education", we offer multiple school-based curriculums, including *Musical in Classroom*, Create *Stereo Modelling*, *Spaceflight Spirit Education, and Campus Tour with Hang Bao*, etc., providing a broad space for students.

Interest: Interest is the basis for exploration. The classes should be vivid and dramatic, so as to arouse students' interest.

Flexibility: It generates from interaction and is full of energy and vitality. In classes, it promotes heart to heart discussion, facilitate the collision of thinking and ideas, and make students think from multiple perspectives to get inspiration and enlightenment; in this way, students can become clever, savvy, and intelligent gradually.

Advocating aesthetics: It is the final objective of classroom teaching under the Time and Space Discovery Education. In the process of classroom teaching, we pay attention to cultivate students' ability to feel, appreciate, show, and create beauty. Meanwhile, advocating aesthetics is able to enhance the sense of responsibility of students and teachers to contribute to society and have great love.

Autonomy: Based on activity themes, students autonomously design activity plans and organize activities to train their own discovery competence.

Openness: The family, society and school combine together to carry out moral education activities; the students walk out of the classroom and into the society to carry out various activities by combining with educational special themes. Open the thoughts of activities, explore knowledge, and cultivate abilities.

Multielement: Forms of activities are multivariant, providing students a good opportunity to study and research, participate in and experience, as well as practice and create.

There are blue sky, rockets towering over the land, and murals on the teaching buildings in the school, displaying the interpretation of life, artistic and technical campus. Different contents are presented on each floor according to different themes, including cosmology and life, arts and humanities, earth and science & technology, containing various elements in ancient and modern times, as well as in Chinese and Western countries, which highlights the philosophy of "Time and Space Discovery Education".

Thematic map in the school hall

Life gallery

Art gallery

Technical gallery

In order to highlight its cultural characteristics, the school has standardized the logo, designed the badge and created the mascots loved by students, and applied the cartoon images to the daily study of students.

Beijing Fengtai District Donggaodi Third Primary School

School badge

Our school badge is the abstraction of Chinese character "three", standing for the characteristics of life, technological and artistic campus. As a whole, it is a triangle, like a rocket which is taking off to the vast universe, representing prosperous Donggaodi No. 3 Primary School that is ready to setting out. Its upper part is the transformation of English letter D which is not only the first letter of "Donggaodi No. 3 Primary School" but also the first letter of

"Discovery". With succinct lines, our school badge is designed with an idea consisted of elements from China and the western, looking vivid and full of technical sense.

The image of mascot is derived from the launch vehicle and is personated, conforming to the characteristics of the student's age; besides, mascot is named by combining with the activities for electing campus star, and the parents of students participate in the naming. The cartoon image is printed on the class brands and reminders. The students name it as "Hang Bao", in which the "Hang" means spaceflight, navigation and so on, requiring courage, strength, wisdom and the like and the "Bao" refers to every child who is the treasure of the family, teachers, motherland and even the world, so every student is "Hang Bao" and will sail under the guide of Time and Space Discovery Education!

Little Doctor in Technology

Proficient in Labor

Learning Scholar

Little Sportsman

Little Messengers of Civilization

Little Art Star

Mascots

5.2.2 Policy Implementation

In the implementation process of school culture scheme, there are four policies for the school to refer to, respectively including the decomposition and division, by stages and themes, meticulous supervision and guidance, and leveraging social capital. All policies are introduced as follows.

5.2.2.1 Decomposition and division

Decomposition and division refer to decomposing the objective and work of culture scheme, and dividing the tasks to responsible persons. The decomposition of objective is the process of decomposing from responsibility objective to work objective, which can be decomposed level by level. The common decomposition and expression methods are tree diagram and fishbone diagram. The culture construction objective will be divided into many task units after being decomposed. The decomposition of work is to decompose the main and deliverable achievement into smaller task units which are easy to manage, from the top to down, thick to thin, till to the smallest manageable units, namely to assign work to everyone.[5] The common work decomposition methods are Work Manual for Positions and Responsibility Chart. Responsibility Chart is the method to show the individual responsibility in finishing the work unit of work decomposition structure in the form of chart. Decomposition and division policies require the clear interpretation of every task, responsible person, members, time and ways to finish, the forms of deliverable achievement, the time for Principal to check, the time for person in charge to report work and the like.

5.2.2.2 By stages and themes

Culture and culture construction are a long-term process, while the school culture scheme is a medium and long-term development plan interpreted and designed systematically, which shall be carried out and conducted by stages and batches orderly to guarantee one construction key point, which can be the weak link or advantageous activities of culture, the newly designed organization structure or ceremonies, etc., for each semester. On the premise of comprehensive consideration, the culture construction can be conducted by themes which coincide exactly with the work area of school, including such themes as management, curriculums, teaching, classes, teachers, students, studies, campus environment, and ceremonies. If the school members have enough energy, they can carry out more work on the basis of fulfilling their own responsibility.

Beijing Fengtai District Fangchengyuan Primary School, one of the project schools, abstracted the "lustrous education" in 2009, striving to make the school become the spiritual home for teachers and students to grow, and make each wall, classroom, and corner be saturated with the cultural, tolerant, invigorative, peaceful, and happy breath of life. Under the leadership of the Principal, Fangchengyuan Primary School took the "small but delicate" and "small but flexible" as the basic construction thoughts, and got primary achievement. Its delicate schoolyard environment and kind home sense have been recognized by teachers and students. The school spent a half year

[5]Yu Shixiong, Winning on Execution [M], Beijing: China Social Sciences Publishing House, 2005: 54.

to extensively disseminate the lustrous philosophy, and then focused on developing the typical logo which can reflect the core value of life education and extensively applying it to the school decoration. The school chose yellow and green as the representative colors of the school and used them to decorate the schoolyard. The green plants stand for peace and harmony, as well as spring and hope; while the light and warm yellow brings happy, hope and wisdom to people. The bell is soft music. The school decorates the school gate with the lustrous as its theme: in consideration of the rich meaning of lustrous in traditional Chinese culture and its characteristic of traditional Chinese painting, Fangchengyuan Primary School can carry out a lustrous traditional Chinese painting competition to collect traditional Chinese paintings that can best reflect its educational characteristics and adopt the best work to decorate its school gate. The next task key point is transferred to the culture construction of curriculums.

Beijing Fengtai District Eighth Primary School, another project school, has successfully combed out and carried out the educational philosophy of "peace and harmonies". With the help of the expert group, it changed weiqi project into weiqi spirit and improved the project characteristics to cultural characteristics. It focused on curriculum culture construction, especially on the design and improvement of the school-based weiqi curriculum on the premise of combing out the philosophy well. Firstly, longitudinally carry out the school-based weiqi curriculum for students from grade one to grade six. For students in low grades, combining with the thinking development characteristics of the pupils in low grades, focus on telling the stories about weiqi to students, namely tell various interesting short stories to them to introduce the production and development of weiqi and let the students to recite classical Chinese poetry related to Weiqi. For students in intermediate grades, focus on the teaching of weiqi skills, namely let students really understand the art of weiqi, to make the students be able to map out strategies and gain a decisive victory in distant, which is the learning of "skills". Meanwhile, the education on etiquettes of weiqi and traditional culture can also be added in the teaching, so as to lay a good foundation for next teaching. For students in senior grades, on the basis of teaching weiqi skills, teachers can primarily teach the students the traditional culture and even the introduction to Chinese philosophy. Mastering weiqi skills is just the first step of weiqi characteristic education. What's more is to cultivate students to understand the "Tao" in "skills", living up to the purpose of not only knowing the skills but also understanding the Tao contained. Secondly, carry out the transverse weiqi education combining with the characteristics of each subject. Weiqi is filled with the glories of traditional Chinese culture, so the teaching of each subject can be integrated into it. For instance, the school has integrated the art and PE into weiqi in the forms of weiqi picture and rubber band gymnastics, which have achieved a good teaching achievement. Moreover, the school-based curriculums for Weiqi and Literature, Weiqi and Math, Weiqi and Astronomy are under positive research and development, among which the school-based curriculum of Weiqi and Literature has been in embryo, and the school-based curriculums for Weiqi and Math and Weiqi and Astronomy have been developed in script. The research and development of such three subjects are based on subject teachers. Through the research and development, teachers really understand the extensive connotation of weiqi and the broad and profound Chinese culture, which is a great challenge and surmounting on their professional standards. Thirdly, hold an annual culture week with weiqi as the theme based on the implementation of characteristic weiqi education, so as to intensively display the achievements of the characteristic weiqi education within this week, and set up a broad platform for students to show their talents. The school can carry out such activity for a week annually and actively disseminate it to the society and the parents of students so as to make the parents and community know about the activity carried out by the school and gain supports from them. On the long run of development, such activity can be blended into the tradition of the school to be carried out annually. Besides, monographs and theses on such activity can be written, and all data can be compacted into disc-recordable for future. Make weiqi culture week one the characteristic activities, and disseminate it actively, so as to let all sectors of the society have a strong understanding of such activities and then accordingly improve the reputation of school in the society.

5.2.2.3 Meticulous supervision and guidance

Good supervision, inspection, and guidance are effective guarantees to implement the school culture scheme; otherwise, any good plans will be a fantasied edifice. Supervision and inspection are a control on the process, which is to find out and correct all actions that violate the school culture construction plan, as well as prevent and correct all deviations and errors in the implementation process of the plan. Only in this way can we guarantee that the school culture construction plan can be implemented better in practice, and all tasks can be carried out well.

The leadership team and the construction team of the school may frequently inspect and guide timely in the form of on-site office. They are required to set up a communication system to ensure a smooth communication and feedback channel. The school can provide several communication and feedback channels to ensure that the suggestions and advices from all teachers and students can be spread to the management of the school in time, such as setting up a principal mailbox and a work complaints book, or conducting face-to face interaction and communication by holding various teachers-students meetings. All members of the school should participate in the culture construction, because everyone is the creator of culture and has the ability to provide different suggestions and advices from their own prospective. In the construction process of school culture, the school has to collect suggestions and advices from all members in time, which, on the one hand, can improve the school culture construction plan and on the other hand, meet the development demands of teachers and students, which is also one of objectives of the school culture construction plan.

5.2.2.4 Leveraging social capital

In the driven model of school culture, the school can improve its strength with the assistance of various external social capitals. For example, take and use the wisdom and reasonable resources from college experts, learn the epistemology and methodology as tools to systematically consider the school development, establish more efficient public relation with the administrative department for education by using the three-party cooperation opportunities to promote the policy level, and carry out activities for students to cultivate a good social adaptability, in which the community and social recourses is always used. For instance, a primary school sets up a school-based curriculum by virtue of the golf course in the community; a middle school designs a scientific simulation laboratory with the help of a student's parent who is an academician of the Chinese Academy of Sciences and Chinese Academy of Engineering; the school fellow of one school provides a huge amount of investment for the school. Meanwhile, the students always give performances or participate in social activities on behalf of their school, etc.

In the implementation of each action plan of the school culture construction plan, especially the design and selection of school motto and badge, the school can extensively collects wisdoms and suggestions from parents, community, and all sectors of society. For instance, the establishment of the Cuiwei education system of Beijing Haidian District Cuiwei Primary School was conducted in this way. The school did two things in determining the cultural philosophy. Firstly, deep survey, participation of experts and clear thinking. With the aid of "school culture construction project" carried out by the

Haidian District Education Commission and Beijing Normal University, the school introduced experts to diagnose the school, invited experts to visit the school for several times to inspect the educational environment, held forums on such levels as principal, intermediate management cadres, parents, and students to conduct "brainstorming", which, through a comprehensive understanding in the school, helped the school examine, know about and go beyond itself as well as provide thoughts and methods to accurately position the school cultural philosophy. Secondly, extensive collection, open media and improvement absorption. With the inheritance of the thinking offered by experts, the school freed its cultural philosophy from the ivory tower, and aroused the enthusiasm of teaching and administrative staff, mobilized social resources available, and collected all wisdoms to offer strategies for the school culture. The school collected the school motto and badge from all sectors of the society, teaching and administrative staff, and students' parents via network, newspapers, and school magazine, to draw more attention, win supports and gain more resources. The school collected more than 200 school mottos and dozens of school badges. Finally, through repeated comparing and screening, cadres of the school selected a school motto made by a parent and a school badge made by Mr. Fan Hongwei. Therefore, the cultural philosophy of "cuiwei education" with "bright virtue and doing the right thing" as the core value pursuit was established preliminarily.

5.3 Result Assessment and Feedback

The school culture construction is a long-term and systematic process that involves all departments of a school. Every department needs to perform its own functions and fulfill the intended goal in scheduled time. The result assessment and feedback refer to the assessment and feedback of the executive process and result of the school culture construction. The difference between the assessment and feedback referred herein and the school cultural assessment described in Chapter 3 lays in that the assessor of the execution result of the school culture scheme for the former one is the school itself and the assessment belongs to self-assessment. The assessment can be conducted at any time and the assessment results mainly serve for the improvement of school culture scheme and behavioral modification and implementation as well as attitude adjustment. The assessment and feedback referred herein mainly mean the specific practices carried out based on the school. The school culture assessment described in Chapter 3 includes the preliminary diagnosis and post effect assessment on the state of school culture development and it is mainly set forth from the angle of macro procedures and steps. The assessors are mainly university experts and the assessment is conducted by others instead of the school. Such assessment will be carried out once a year or once for years with the involvement of others. The assessment results have two purposes of use: one is to provide the school with feedback information to modify behaviors or adjust schemes; the other is to serve for study, modify the driving model of school culture, and adjust the attitude and behavior of the researchers who are engaged in improving the school. Definitely, the assessment methods, techniques and instruments can be shared in both self-assessment and external assessment.

5.3.1 Content of Assessment and Feedback

In this link, the executive process, link and system of school culture scheme, as well as the execution and implementation results, effects and efficiency of the scheme, shall be evaluated, and the execution results of school culture sheme shall be exchanged and fed back among all parties.

5.3.2 Strategy for Assessment and Feedback

From the angle of cultural management, in the executive process of school culture scheme and the assessment and feedback process of results, the school can take the strategies as follows for reference.

5.3.2.1 Convene assessment conferences

In the early yeas of carrying out a systematic school culture construction, the assessment mobilization meeting and training and mentoring meeting involving all faculty must be convened to make clear the meaning of self-assessment and assessment by others, conduct technical training and psychological mobilization. The concrete content of assessment methods and techniques of school is seen in Chapter 3.

5.3.2.2 Make self-assessments

In the process of school culture improvement and construction, self-assessment is needed everywhere and at anytime. Self-assessment is often carried through in the following situations: (1) in the situation where comparing with other schools. When going out on a tour of investigation or communicating with brother schools, or listening to the feedback of teachers on work in some certain aspects of other schools, or reading some information of other schools, the principal will naturally make comparisons between his school and others'. Grasp different aspects and then make analyses on the organizational culture atmosphere, education and teaching activity, and relevant systems or campus environment construction; thus, it gets into a self-diagnostic state of organizational culture. (2) in the situation where to resolve the practical difficulties in work. Difficulties are ubiquitous in work. Although the natures of those difficulties are different, the process of resolving difficulties can reflect the characteristics of organizational culture. Analyze the organizational factors leading to difficulty, solutions to the difficulty and process of resolving difficulty and then purposefully make analyses by combining the attitude and behavior of facing with the difficulty with the internal factors which lead to the attitude and behavior; thus, it gets into a self-diagnostic state of organizational culture. (3) in the situation that after summing up the successful experience. After one stage of work is finished, the school often makes a routine summary. In the past, such summary usually confined to the gain and loss of some certain methods and ignore the internal background of individual behavior. But the internal background precisely reveal the teachers' approach to perceive problems, thoughts to analyze problems and value orientation in solving problems under the influence of organizational culture. If we can purposely analyze the individual behavior when summing up the experience, we will get into a self-diagnostic state of organizational culture.

For professional self-assessment on execution results of school culture scheme, the school culture assessment mode and instruments in Chapter 3 can be taken as reference. It is an assessment by others and self-assessment carried out from the angle of research and can evaluate the execution result of culture scheme by contrasting the change of pretest data with the posttest data. What is highlighted in Chapter 3 is the external assessment process of school and Zhang Wenqing comes up with the general steps for self-diagnosis of school culture for the application in self-assessment of school.[6] The conception of "self-assessment" and "self-diagnosis" referred herein shall be regarded as the same. The steps are as follows:

1. Design self-diagnostic program. The concrete content of the scheme includes: diagnostic questions, diagnostic purposes, diagnostic content, diagnostic objects, concrete operations of diagnostic "six-step method",

[6]Zhang Wenqing. Entry Point, Method and Steps for Self Diagnosis of School Culture [J]. Primary and Secondary School Administration, 2004 (7): 1113.

diagnostic methods, basic requirement of acquiring information, diagnostic time and procedure, time of finishing the diagnostic report, etc.

2. Carry out self-diagnostic scheme. It mainly refers to executing the diagnostic "six-step method" which is the core method for carrying out self diagnosis. "Six-step method" is not simply about six questions and six answers but refers to that the investigator and respondent gradually go deep in the organizational culture at the back of the educational phenomenon step by step in the investigation process until the organization is reformed.

   "Six-step method" treats both the respondent and investigator as the subject of investigation. The respondents, by pondering over his or her behavior, assist the investigator in entering in the layer of organizational culture to perceive problems and help the investigator to achieve the diagnostic purpose via diagnostic procedure. The respondent is not the diagnostic object nor the provider of materials and information, but the master of studying problems in the process of diagnosis.

   "Six-step method" cannot achieve the purpose unless by "change". "Six-step method" requires the investigator to clearly know about the purpose of every step of it and meanwhile fully understand the internal relations between every two steps, so as to form an "inner investigation structure". With the guidance of such structure, according to the problem of entry point and combining with the characteristics of the respondent, the investigator works out the "contact topic", "discussable topic", etc. and hence form an "investigation structure for application" in the investigation process. The investigation structure applied in actual investigation is always in a state of "depth talks" so as to achieve the purpose and task of "inner investigation structure" in the end. The interviewer and interviewee need to make a response and analysis on these problems raised from the six-step method: what is the current situation? Why does the current situation happen? What's your feeling in the current situation? What do you want to do? What do you plan to set about? What support do you need from the organization?

3. Sort and analyze material. In the diagnostic procedure, continuously sift and observe the "depth description" which reflects the function of organizational culture, extract the words and behavioral styles outpoured unconsciously from a majority of people in the interview and symposium and conclude their common concept and behavior from those words and behavioral styles. With the material accumulation of "depth description"

of phenomena in diagnostic procedure, there's a basic material for the diagnostic report.

Four aspects of work shall be done to the sorted materials d: (1) learn the latest theories concerned with the basis of the content of diagnosis, and use new concepts to work out the ideal organization mode of the relevant aspect and put forward basic thinking of an ideal organize. (2) Find the differences between the subjectivity and objectivity. Rethink the organization subjectively understood by the administrator, the organization specified in the school text, the organization in actual running and the organization in the subjective wish of subordinates; "hang" up the different organizations to find out their differences. (3) Sort out the forming process of the organization subjectively understood by the administrator, the organization specified in the school text, the organization in actual running and the organization in the subjective wish of subordinates separately; then, find the entry point for organizational reform based on it. (4) In accordance with the consideration of ideal organization and the entry point for organizational reform, explicit regulations for new organization in written form. In regulations, the feasibility of actual running shall be especially concerned and some extent of openness shall be remained. To make sure the clarity and conciseness of the material recorded in diagnostic procedures, it is suggested to adopt a chart to record in diagnosis.

4. Write a self-diagnostic report of school culture. The main content of self-diagnostic report of school culture covers: the main behavior phenomena of diagnosis; the observation and description of behavior phenomena by administrator; the content and conclusion of interview or symposium of teachers, students, parents and administrators; "hang" the organization subjectively understood by the administrator, the organization specified in the school text, the organization in actual running and the organization in the subjective wish of subordinate and find the differences between them; make an analysis on the cause of generation of the behavior and organizational culture; on the basis refereed hereinbefore, put forward the entry point for organizational reform and concrete management arrangement.

What needs to be stated is that the "six-step method" stated herein shares the same logic and principle with the content of Chapter 3. In the project schools where we have carried out experiments, the assessment modes, programmes, methods and technologies provided by us can be adopted.

5.3.2.3 Collect the data of assessment by others

Assessment by others is a kind of external assessment, including the assessment on school by governmental supervisory authorities and also the collection of assessment data from parents and community members by school consciously.

The first is to collect the data from governmental supervisory authorities. In the culture construction process, it is inevitable for the school to accept the supervision and assessment from the government and such assessment may be an assessment on cultural topic, an integral assessment or an assessment on other topics. No matter what the focus of the assessment is, the school culture improvement and construction will be involved in. The school can carefully collect the assessment conclusion and opinions every time from the supervisory authorities and then make comparisons.

The second is to collect the data from various media. When the educational administrative officer at all levels come to the school to inspect the work, the school can make a video and take photos purposely and the speech can also be recorded and worked out into written form or light disk to keep these data completely. In different development stages of the school, the theses or special reports and other written media data published on various newspapers and periodicals also need to be arranged by specially-assigned people and can be sorted out to compile in book form. The report and promotional material, about the students or courses, from broadcast, television and internet also need to be collected carefully. The school can collect and save these materials into the school history museum or school museum after arrangement or make a further data analysis on them to scientifically evaluate the cultural achievements of school through discourse analysis and physical analysis. For example, Beijing Fengtai District Donggaodi Third Primary School summarizes the effect of “Time and Space Discovery Education” like this: through school culture construction, the “Time and Space Discovery Education” of our school has enjoyed a popular support and gained the identity of teachers, students, parents and the society; moreover, we often exchange experience with other schools on the municipal and district working conferences. Our school has been rated as the educational brand specialist school in Beijing, scientific research advanced school in Beijing, high-quality school for fully carrying through education for all-round development in Fengtai District, etc. The television station reported the school running situation of our school: special fund for teaching and research development of Chinese Foundation for Teacher Development, in special cooperation with China Educational Television, held the “My Dream, China Dream”–TV Show of Teachers and Students’

Dreams in Elite School Nationwide for the purpose of cultivating the imagination and creativity of students, and encouraging the students to build up the correct ideal and belief. China Educational Television Channel 1 broadcasted the dreams of the teachers and students of our school on October 6, 2010. On April 29, 2011, the programme "Chinese TV Documentary" of China Educational Television Channel 3 broadcasted the cultural characteristics of our school, i.e. "Space-Time Exploration Fosters New Star".

The third is to collect the assessment data from parents and communities. In the culture construction process, the school needs to organize the interested groups related to evaluate the executive process and results of school culture scheme by stages and collect the process data and result data. The parent representative and community representative symposiums shall be held in every semester to collect data and gain information by means of rating scale and questionnaire, as well as question-and-answer form and symposium. For example, the questionnaire and satisfaction survey on the identity of school culture of parents are frequently used. Parents and the community will assess whether the actions of school are in conformity with the cultural spirit publicized in accordance with the behaviors of students in family and community activities and then show their attitude and opinions accordingly.

5.3.2.4 Feed back in regular stage

The school culture construction group shall regularly organize the school culture construction symposium on which each functional department or sub-scheme action group shall submit the staged working reports. The report should cover a detailed statement on the staged achievements, problems, difficulties and practical experience in the constructing of school culture within department. Then, according to the feedback information from all departments, repeated discussions should be conducted to estimate whether the current completion status conforms to the anticipatory goal of school culture improvement and construction, so as to take solutions directing at the problems. With regard to the difficulties existed, the school members should give advice and suggestions to solve them jointly; the experience and lessons concluded in the process of construction by all functional departments should also be taken out for sharing to serve as a warning and avoid detours.

5.3.2.5 Feeding back results by school

With respect to the results of school culture improvement and construction in one stage or a period, regardless of the results displayed by quantitative data or qualitative data, the school should feed back the results after making full preparations. The first is to convene the feedback conference and improvement conference. The principal should report the assessment results, construction results and improvement program to the faculty in person; besides, suggestions and opinions can also be fed back to all action groups on conference or after conference to be prepared for the application in improvement program and actions. The second is to feed back the results and achievements to the parents of students by making use of the parents meeting and collect opinions from parents. Photos of some classic achievements can be contained in the school culture manual for sharing with the students and parents. The third is to publicize the results on campus network to make more people know them and gain more popularity as well as enhance the school reputation.

5.3.2.6 Feeding back to school

The school will accept the individual supervision or integrated supervision assessment by the government and the school culture improvement and construction results would also be checked inevitably. The governmental supervisory authorities need to feed back the supervision results to the school in time. After one year of the implementation of culture scheme, it is necessary for the university experts to pay a return visit. The school should invite the experts to have an earnest talk, reviewing the achievements, speaking glowingly of the experience, pointing out the confusion, holding a joint discussion of countermeasures and adjusting and deepening the school culture construction program. The school should fulfill the adjustment of school culture scheme within a month and then continue to execute and carry through the scheme. The school should conduct the abovementioned processes circularly, so as to improve the school culture construction spirally. What follows in the passage is the return visit record written by Dr. Xu Zhiyong, a member of the project group, to me. Dr. Xu served as the contact person of this school from 2008 to 2009.

Thoughts on the Return Visit to Fangguyuan Primary School January 21, 2010

Ms. Zhang,

Hello!

Yesterday morning, I took the postgraduate Liu Ruijie to go back to Fangguyuan Primary School. Our project group, on the basis of summing up the school running experience, came up with the idea and cultural system of "entire faculty service and precision management" for Fangguyuan School in last May. Through one semester's implementation, Fangguyuan Primary School has made a continuous process in its culture construction.

Guided by the school culture of "entire faculty service and precision management", the school systematically summerized the school running history and reality and has planned the implementation of school culture in the future. The Principal Shen Ruizhi, on the district on-site scientific research meeting on October 23 of last year, school-running supervision meeting on November 17 and the periodical summing-up meeting on standardization construction of Fengtai Primary School at the end of November, had presentations under the guidance of this culture concept, and hence leading to a good effect, which manifests in two aspects:

First, the leading supervisory expert Liu Zhanliang said, "I have never seen a principal who figures out the issues of school that clearly!" On the acceptance of appraisal, the Fengtai Board of Education wrote, "'entire faculty service and precision management' has seeped into every piece of work in school with clear goas, highlighted key point, explicit responsibility assignment, effective measures, strong operability and detectability as well as high awareness rate among faculties."

Second, on January 7 in the new year, the Director of Fengtai Board of Education, supervisory cadres and over 120 principals and leaders came to Fangguyuan Primary School again (because of the extremely success of the first time) and held a "on-site meeting on supervision work" and carried through the scientific outlook on development. Principal Shen made a report centering on the concept of "entire faculty service and precision management", which won a round of applause. The participants of the meeting were reluctant to leave the meeting place after staying for a long time.

At present, through a long-time thinking, Fangguyuan Primary School makes a further explanation to the concept of management style of the Principal, behavior culture of cadres, behavior culture of teachers, etc. and now they are more performable. Next, Principal Shen gives me a task which is to assist the school in further refining, executing and humanizing the actual operation and theory enhancement to manage the school with system and culture other than by human. We will get in touch with you in the holiday.

I think Fangguyuan Primary School is a school with strong cultural systems thinking and performability and can be studied as a typical case.

. . .

Hope you all the best!

Xu Zhiyong

5.3.2.7 Solidify achievements

The execution and implementation of school culture scheme is a process which researches and operates concurrently and a process with syntonic academic thinking and behavioral thinking. All process data and research data can be sorted into multiple forms of achievements. The driving mode of school culture advocates, encourages and helps the school to solidify the achievements and record the efforts made and footprints left in the development of school culture. There are three strategies for solidifying the achievement listed below.

The first strategy is to work out the report on school culture construction process, assessment report and other reports, keep and record the research evidences and processes. The second strategy is to publish research theses or research reports on educational newspapers and periodicals. In our project, the university group, primary and secondary school principals, teachers, teaching and research staff have published tens of theses and taken a professional effect as well as intensified the strength of promoting the culture construction achievements of project schools. For example, the thesis of thinking on the educational concept and practice of childlike innocence compiled by the Principal Song Jidong of the project school, Primary School Affiliated to Capital Normal University, and the thesis on development education concept and practice compiled by the Principal Liu Fang of the Primary School Affiliated to Chinese Academy of Agricultural Sciences published on the second issue and sixth issue of *Primary and Secondary School Administration* of 2010 have a wide influence. The theses of the principals of ten project schools in Fengtai District have been published on *Teacher's Journal*. The third strategy is to publish monograph. Provided that the school culture scheme has been carried through for over three years, the summary can be made and a monograph, featuring technicality and research on school culture construction, can be compiled under the general editorship of the principal. The university group shall guarantee a strict standard over the book name and content structure and take the advantage of cultural capital to make contact with famous publishing houses for project schools. Publishing the monograph featured with research-oriented characteristics is not an easy thing for the project schools. First, the school cultural monograph compilation team shall be established under the leadership of the principal. The outline and the first draft need to be discussed with the university group for numerous times and, after the completion of final draft, it requires to communicate with the editor of publishing houses. The book named *Defend the Childhood: Educational Concept and Practice of Childlike Innocence*, which was under the editorship of the Principal Song Jidong of the Primary School Affiliated to Capital Normal University, has been published in August 2011 by Beijing Normal University Press. This school is the first one which has published the school cultural monograph so far among the project schools. What follows hereinafter is the work record of first draft revision by the project group.

Record of Draft Revision Symposium of the Primary School Affiliated to Capital Normal University.

Time: 8–12 o'clock, January 12, 2011.

Place: conference room of the Primary School Affiliated to Capital Normal University.

Participants: Zhang Dongqiao, Ma Tusheng, Yu Kai, Xu Zhiyong, Yu Qingchen, Zhao Shuxian, Principal Song Jidong, Secretary Shi, et al, totalling 10.

Host by: Principal Song, Ms. Zhang.

Proceeding:

Principal Song makes an introduction (omitted), the experts present opinions (omitted). The summative suggestions of Professor Zhang Dongjiao are as follows.

About the book name. The present name "Childlike Innocence Makes the Campus a Pure Land" sounds like a ragged verse. I advise to adopt "Pure Land for Childlike Innocence" or "Pure Land Childlike Innocence" (Beijing Normal University Press finally published it with the name of "Defend the Childhood: Educational Concept and Practice of Childlike Innocence").

About the orientation and altitude. The opinions of them sound divergent but reach the same goal by different means in fact. The divergence is the contradiction between the perceptual and the rational and the contradiction between the expression of ideological system on school management and practice system and the readers' preferences. Our opinions are for reference only and the school needs to make decision by itself. In my view, the altitude of academic nature and research don't contradict the selection of specific narration mode. The narration mode will not impact the research nature if the altitude is built up. Formal style and euphemism can be used concurrently, but it would be better if only one narration mode is selected. The highland of childlike innocence educational theory shall be grabbed and the methodology and epistemology shall be controlled. The following several problems about altitude need to be well handled: the relationship between the freedom and social regulation pursued by childlike innocence education; the relationship between the innocence and purity and the sophistication; the relationship between the virtue and the dominant values of the times. These deliberations should run through the entire book, serving as the point of view of perception and philosophy standpoint. As with the narration mode, it's not suggested to adopt the style of writing by separating the theory and practice operation, for it is easy to make the thesis abstract and complicated. Therefore, I suggest applying the conceptual disintegration way to carry out several themes and each theme shall be supported by key event and typical case.

Some suggestions for structure adjustment. Formal structure adjustment: add two parts of introduction and conclusion, with the main body be consisted of six to eight chapters. Adjustment in content structure: a chapter about students shall be added, such as the lively childhood or child's voice; a chapter about childlike innocence shall be added. Then the whole book would become more rhythmed. The whole book has 150,000 words. The serial number in the form of "I" or "1" is better than the chapter or section. Serial number shall be less used. The existing titles are usable but they should be consistent in the way of childlike innocence education, childlike innocence management, childlike innocence teaching, childlike innocence courses, childlike innocence virtue, teachers with childlike innocence, kids with childlike innocence, childlike innocence environment, childlike innocence culture and the flexible narration modes can be added in it by consideration. For example, childlike innocence environment: there's a pure land here; teachers with childlike innocence: such a person is deserved to be the teacher of kids. Direct and indirect quotation should be noted with footnotes according to standard format and the sources of the quotation shall be accurate to the pages.

The personal suggestions to content structure are as follows.

Introduction: who break the wings of the swan? Use this question to lead to the reason why this book is written and how to write it.

1. Childlike innocence education. This chapter has about 30,000 words and is written by the Principal in person. The following aspects shall be

covered: what is the childlike innocence? What is childlike innocence education? What are ideological system and practice system? The content can be displayed with charts or tables. What's the cause of childlike innocence education? Theoretical thinking, school tradition, joint efforts, etc. are also included.

2. Childlike innocence management.
3. Childlike innocence teaching.
4. Childlike innocence curriculumns.
5. Childlike innocence virtue.
6. Teachers with childlike innocence.
7. Kids with childlike innocence.
8. Childlike innocence environment.
9. Childlike innocence culture.

Conclusion: give you a pair of wings. These suggestions are only for reference for school.

What calls for attention is that the foregoing school cultural practice promotion is only one of the units and stages of school cultural management and construction. The school culture construction process under the guidance of planning program of school culture development is also a process of continuously digging and improving the core idea of school culture. Many new and valuable thoughts produced in practice need to be extracted and supplemented to the core ideological system of school culture, for it is beneficial to promote the continuous enrichment and a realistic ideological system. Then, the promotion of school cultural practice enters into a new stage, namely to carry out a new round of propaganda and identification of the adjusted and upgraded school culture, implementation and execution of program and the assessment and feedback of results. Thereby, by that means, a spirally-escalated structure of school culture construction is formed subsequently and the school culture construction and management will be continuously promoted to a higher level.

6

School Culture Conflict and Solidarity

School culture improvement is a rational process and school culture study is a process of thinking and operating. Although we say that "every school has its own culture", we must make clear of what requirements the school culture has before carrying out the school culture improvement practice system. Only by grasping the nature of these requirements can we master the general principles of school culture improvement and construction. Once the key link is grasped, everything else falls into place. Similarly, once the general principles are grasped, school culture improvement will come into good effect.

This book develops by centering on the concept of school culture and school culture improvement and it takes such logic: abstract-specific-abstract. The first three chapters answer the following questions from the angle of abstract theory: What is school culture? What is school culture improvement? What is school culture driving model? What is school culture assessment? The fourth and the fifth chapter solve these problems from the angle of concrete operations: How to make plans for school culture? How to put school culture plans into practice in order to promote cultural practices? Standing on the position of sociology to rethink our study and what was happening in front of us, this chapter is to explore the occurrence mechanism of school culture, find academic basis of school culture improvement practices described earlier and reconsider and answer these three questions by abstracting representational school and its cultural practices out and returning to abstract theory level: What is the rationality and legality of the school culture? How does the school culture happen? What is the purpose of school culture improvement?

Taking a broad view of practices and reflective study, we may discover: Rationality and legality is the source and premise to support school culture and make its practice carry out in full swing. The growing and strengthening of school culture is the process of continuous spiral from conflict to solidarity and move on to a new conflict and then to a new solidarity. Grasp the mechanism of school culture, foresee and timely resolve possible problems during the school culture improvement and construction process, accumulate energy, enrich and strengthen culture data points and occupy a favorable position in the school culture spectrum, so as to lead the school to make progress and develop, as well as make it be distinguished from its kind. In sociological sense, school culture solidarity and culture prosperity is the real purpose of school culture improvement.

6.1 Rationality and Legality of School Culture

As a kind of educational organization, school is different from other social organizations; also, school culture, with its own distinctive features, is different from other organizational culture in society. Unique features of the school culture are its rationality and legality.

6.1.1 The Rationality of the School Culture

Max Weber distinguishes and judges whether an action is rational and the degree of rationality through the concept of rationality while understanding social actions. Hereon, we discuss the rationality feature of school culture by virtue of the concept of rationality.

Rationality can be divided into instrumental rationality and value rationality. Instrumental rationality refers to regarding circumstances of external things and expectation of others as targets. Value rationality refers to regarding the pure faith of a particular values or behavior as the target regardless of its success or failure. Any blind, no thinking and taking personal emotion and habitual for guidance is irrational. Traditional factors and emotional factors belong to the irrational factors, but they can be converted into rational factors.

6.1.1.1 Instrumental rationality of school culture

Instrumental rationality mainly considers the adaptation of a certain education activity with its surrounding environment and conditions, and also the ability and possibility for some available means to achieve a particular goal. Instrumental rationality insists the neutrality of value and its standard is the objective condition and environment. Its result is to obtain greater utility. Obtain greater output with less investment.[1] Those schools that extremely pay attention to instrumental rationality usually do everything very well, but they lack unified thought leading and the wavelength of the school culture spectrum length is different. Generally, these schools are able to complete the assignments of the administrative department for education, but their personality development is short, connotation is not enough and creativity is relatively weaker.

Instrumental rationality actually emphasizes the efficiency problems in the course of school culture improvement and construction, namely the school's actual situations need to consider while carrying out school culture improvement practice, operational problems of school culture plan and the problem of results expectation while carrying out school culture plan.

Any school culture construction and management should start from the actual conditions of the school; therefore, it needs joint efforts from principals and teachers.

Principals should have a comprehensive consideration for the school's overall situation, especially think about reasons that induce changes in the school culture, and make teachers and students of the school understand the importance and necessity of culture management and construction.

Most importantly, the principal needs to make predictable assessment for the result of culture management and construction and should not ignore existing cultural accumulation and cultural reality of the school during the process. In addition, the principal also should consider proper ways to communicate and talk with teachers, plans that can be accepted and willingly participated in by teachers, improvements that can help the existing culture make innovation and breakthrough, etc.

For teachers and students in the school, school culture plan is likely to become a reality only when they have demands and recognition for school culture and they can foresee actual benefits they can get after the start of new culture management practice. Besides, teachers usually will depend on

[1] Xie Weihe. Sociological Analysis of Educational Activities: a Kind of Study of Educational Sociology [M]. Beijing: Educational Science Publishing House, 2007: 156–157.

the principal and prestigious teachers' attitude toward the reform of and their energy of investment. In a word, in the course of carrying out school culture improvement and school culture plans, we must make be clear what teachers really need and what kind of culture can promote better development for the school, teachers and students.

6.1.1.2 Value rationality of the school culture

Value rationality emphasizes the constancy for target value and it doesn't care practical results of the action but its process. Value rationality always makes action develop towards some absolute and eternal goals. Goals of objective rationality are changeable and the objective rationality always focuses on the choice of effective means.[2]

Value rationality is the action carried out and done based on certain subjective standard and believes that some behaviors have unconditional and exclusive value regardless of the conditions and consequences. It emphasizes whether education activities conform to certain philosophies and criterions rather than considering the consequences. Its feature is emphasizing the rationality of its goal and specific value. It can be the value of the ethics and moral, that of some popular and common norms and principles for social or education behavior, and also that in politics, etc. But these values are all rational but not emotional.[3]

Schools that extremely pursue the value rationality usually have a very good thought and school-running pursuit. But they may don't care about the outcome for concrete affairs. They will do everything and think doing things like this is good for the realization of school values, but they wouldn't stop to think and evaluate the concrete affairs or improve management experience. Such schools usually have a lot of passion in doing things, but can't reach the expected results and lack the habit of summarizing experience. Therefore, the rationality of school culture must effectively combine the two.

Namely, it should not only pay attention to the value and rationality, but also pay attention to objective rationality. In reality, there are a few schools that extremely pursue value rationality and instrumental rationality, but most schools are somewhere in between.

Education is the study activity that teaches people to learn to be good and also a kind of social activity to make people healthy. The function of education

---

[2]Hou Junsheng. Western Sociology Theory Tutorial [M]. Tianjin: Nankai University Press, 2001: 116–118.

[3]Xie Weihe. Sociological Analysis of Education Activities: a Kind of Sociological Research [M]. Beijing: Educational Science Publishing House, 2007: 157–158.

is to cultivate various talents for the society and make humans play a main role in promoting social development.[4] From this perspective, better cultivating talents is embodied in the objective rationality and cultivating better talents is embodied in the value rationality. Only by combining these two can we better realize the development of both society and humans. It is the same with school culture improvement. We should not only consider the mission of value that we shouldered, but also consider the mission of efficiency.

The value choice of school culture in general can be divided into four levels, and they respectively are basic values of human beings, social dominant values, excellent traditional values and occupational values.

Basic values of human beings are some basic value quality cherished commonly in different cultural traditions, such as equality, justice, mercy, peace, integrity, solidarity, tolerance, compassionating the young and weak, etc. Social dominant values are also known as the core values of society and they refer to some value principles recognized by each social class and ethnic group. Compared with basic values of human beings, social dominant values are the bond of mutual recognition, harmonious coexistence and joint development among different social classes and ethnic groups within a country, nation or region, such as justice, harmony, etc. Traditional values are unique value culture of different nationalities formed in the course of survival and development. Occupational values are the embodiment of basic values of human beings in the field of occupational activities, which are consistent with social dominant values and also have their unique occupational ethic traditions,[5] for example, teacher's ethics belong to the educational and ethical norms.

In the course of determining the value rationality of school culture, we should take the above different levels of value pursuit into account. After being integrated and blended, they will become the value pursuit of the school and be embodied as the core value system of the school, namely the school spirit culture system. For all the case schools listed above, the determination of school culture's core value system all follows the guidance of rationality. When schools are determining cultural value rationality, they are facing numerous value choices and they may feel dazzled especially when faced with different levels of value choices. In fact, there are no differences between high and low and advantages and disadvantages whichever the level of value is. As long as

[4]Huang Ji. The Nature of Education [J]. Chinese journal of education, 2008(9): 1–4.

[5]Shi Zhongying. The Mission of the Age of Value Education [J]. China Education Press Agency, 2009(1): 18–20.

we are in line with specific situations of the school, take it as the starting point and the ultimate aim of each school activities and carry it out and implement it into all kinds of regulations and activities of the school.

6.1.2 Legality of School Culture

Legality means a kind of property and condition in which social orders and authority are voluntarily approved and obeyed.[6] Legality is generally used in political order and rights.[7] As the specialized educational institution, the school would be influenced by social politics, economy and culture, and is closely connected with external organizations. Meanwhile, the school is a relatively independent organization, which has certain binding force and control on internal member and also requires the internal members to approve and obey the school. Therefore, legality of school culture is a kind of property and condition in which the school culture conforms to social and cultural norms in external, has authority in internal, and can be voluntarily approved and obeyed.

The ultimate aim of leadership and management of school culture is to let all school members voluntarily accept and approve school culture and forms a unified cultural identity and behavioral style. Therefore, school culture improvement is also the process of legitimizing school culture, which also requires legality in school culture. Legality of school culture includes two important aspects. From the perspective of the relationship between the school and external parties, school culture should conform to national laws and regulations and requirements of external environment. From the perspective of the management of the school and its internal members, school culture should be accepted and approved by internal members.

6.1.2.1 School culture conformed to the requirements of the law and environment

The binding "laws" for schools include national laws, administrative laws and regulations, as well as local laws and regulations. The premise of legality of school culture is that both the school and the personnel qualification of the school are legal. On this basis, the spirit culture, system culture, behavioral culture and substance culture of the school are legal.

---

[6]Yu Keping. Governance and Good Governance [M]. Beijing: Social Sciences Academic Press (CHINA), 2000: 9.

[7]Liu Fuxing. Value Analysis to Educational Policy [M]. Beijing: Educational Science Publishing House, 2003: 47.

The legality of school is mainly reflected in the qualification of the main body that runs the school and the provisions that the school conforms to school-running conditions.

The school must be in possession of the subject qualification of running a school, namely, the ability to operate a school. Any legal person or organization, such as the nation, collective, social organization, group and citizen, can be the school-running main body. Running a school is not for profit; school-running person should not be a person who is deprived of political rights, condemned to imprisonment, or undergoing the abovementioned sentences, but be a person with full capacity of civil conduct.

Conditions for running a school include organizational structure and regulations, such as explicit school-running objective, enrollment target, duration of study, curriculum, teaching qualification, graduation destination, funding source and charging standard, qualified teachers, teaching sites and facilities which conform to the standard set, and necessary school funding and the stable funding source.[8]

There is legality of the school staff. The staff involved in the school mainly includes the principal, teachers and students. In the laws related to teachers and students, in addition to the laws (such as the constitution and labor law) that every citizen should obey, there are also specialized laws related to teachers and students (such as the Teacher's Law, the Law of Protection Minors, and Law on Compulsory Education). Therefore, the management and construction of school culture must conform to those requirements of laws. The school and its legal representative must have competent policy level.

Teachers are those professionals who perform education and teaching responsibilities. Teachers' responsibilities are to impart knowledge and educate people to improve the national quality, obey laws and rules and be a model for others, perform teacher's contract and complete teaching tasks, cultivate students and carry out social activities, care about students and respect student's personality, prevent others from violating student's legitimate rights, and continually improve teachers' quality.[9] At the same time, teachers are also entitled to some rights, such as the right of educating and teaching, academic freedom, the right to guide and manage students, the right to get reward, democratic management right, right to engage in advanced studies, and other rights.

---

[8]Zhang Weiping, Shi Lianhai, Educational Laws [M]. Beijing: People's Education Press, 2008: 161–163.

[9]Ditto, page 207 to 220.

On the one hand, students are entitled to fundamental rights provided by the constitution, such as the right of personality, right of health, right of reputation, property right and right to education. On the other hand, they are also entitled to specific education rights stipulated in the education law, mainly including the rights to using educational resources, the rights to acquire education aid, the rights to get fair evaluation and the rights to appeal and suit, as well as various rights stipulated in the Law of Protection Minors.[10]

There is legality of the school culture. The legality of school culture is meaningful only on the premise that educational institution and school staff are legitimate. Legality of school culture is reflected in the value pursuit of the school, institutional regulations and behaviors and activities that conform to laws and regulations. All of these are the source and premise that the school culture gets legality. In other words, management procedures and specific operations of the school culture discussed in the former chapters can only be operated on the premise of legality. Thus, conforming to laws and regulations is the lowest requirement and limitation for school culture, based on which the consecutive, mature and great school culture can be formed. Legality of school culture is mainly reflected in school value pursuit and that its school-running philosophy cannot violate legal provisions, school material culture construction cannot do harm to the health of teachers and students, various rules and regulations in the school cannot violate the rights of teachers and students, and various school activities cannot violate relevant laws and regulations.

School culture should conform to not only the requirements of laws, but also the requirements of surroundings. Surroundings include concrete environment and general environment. Concrete environment refers to the group or individuals that are directly related to the school, such as educational administrative organization, parents, community, schoolfellows and the media. General environment refers to economy, politics, social culture, technology and international environment. School culture should conform to the development tendency of the surrounding to some extent.

6.1.2.2 Authority of the school culture

Authority and constraint of the school culture is mainly showed in the influence of the school culture on teachers and students. Such influence is restricted by the coordination degree of elements of school culture itself. Authority of school culture is mainly showed in its reasonable and proper

[10]Ditto, page 246.

source; that the school culture is approved by all members; and that it's necessary to have strict measures to guarantee the authority of the school culture.

There are various sources for the school culture, including the cultural traits formed at the beginning of the foundation of the school, and new changes generated due to the occurrence of important events in the development process of the school, such as the changes made to response to new national policies, new educational views proposed by the new principal, and new thoughts and situations brought by newcomers. However, transforming those new changes into school culture needs a process, in which all school staff and members are needed to participate in to discuss. Once the school culture becomes the carrier of school spiritual pursuit, system culture, behavioral culture or materials, it will win certain authority, and become the striving direction or the regulation to be conformed for all members.

In addition to the reasonable source of school culture, the recognition degree from school staff and students is also an important reflection of the authority of school culture. The comprehension on school culture, especially on school core value, will directly influence school development goals and the striving direction of teachers and students. Therefore, the cognition and comprehension on school culture from teachers and students is one of important reflections that show the function of authority of school culture. The generation and acceptance of school culture are both processes that need some time. The acceptance degree of new school culture depends on its superiority, adaptation, complexity, feasibility, distinctions, and traits of the proposer. Only when the school culture is accepted and approved, can its authority be better carried out. For the case schools which are illustrated in the former chapters, the reason why their school culture can be implemented so effectively and successfully is just because the legality and rationality of school culture. This is the important reason why we emphasized the wide participation of the school staff and students and the school executive force. In the drive model of school culture, over ten schools which participate in this improvement are all of these characteristics.

One of important parts of school culture improvement is to determine how to make sure the school staff and students approve and practice school culture. Therefore, establishing measures to make sure the implementation of school culture is particularly important. The principal and leaders should set an example and play a leading role for others in this process. It also needs to timely make feedback about the behavior of teachers and students. See the contents of Chapter 4 and Chapter 5 for details.

School culture improvement should first make sure the rationality and legality of school culture. Only when the school culture conforms to requirements of laws and reflects value appeal of humans, society, nation, group, etc., can the school culture be scientific and promising. Only instrumental rationality of school culture and sufficient authority can make sure the true implementation of school culture, and really become the strong driving force of school development.

6.2 Conflict-Solidarity Spiral System of School Culture

The theory of school culture spectrum explains the expression mechanism of culture result of individuals and groups in school, but it doesn't disclose the internal formation mechanism of school culture. From the occurrence mechanism, the ideal process of school culture from the generation, development and strengthening to the formation of a set of culture system with clear logic that covers core value system and practice system is a rising process in spirals changing from conflict to integration, and from weak to strong, which is called the rising spiral state of the generation and development of school culture. Of course, the state of declining spiral from strong to weak or the state of maintaining stable also exist in the real life. What we are discussing in this section is the rising spiral state. Since the foundation of a school, culture begins to take root and grow there. Such growing may be spontaneous and unconscious, or may be conscious. In the development process of a school, conflicts may appear between the school and outside world, among various groups in the school, and even inside the groups. Along with the settlement of conflicts, school culture will reach a state of solidarity. Meanwhile, in the process from conflict to solidarity, school culture also gradually takes shape and plays its role in it. The management of school culture is the process that the school consciously manages the conflict and then reaches solidarity, which is the process of rising spiral.

The school organization is in society, so it has countless ties with other systems in the society. What stimulates the development of school culture is the change and development of various contradictions in school. These contradictions are mainly reflected by conflicts and solidarity between two parties of the contradiction.

6.2.1 School Culture Conflict

Conflict is a kind of process, which begins when one party feels the other party has adverse effect or will have adverse effect on it.[11] From this definition, it indicates that the cause of conflict lies in the antagonism and conflict between different interest parties. From the perspective of the relation between school and society, the school, as an important constituent part of the society, has its distinctive characteristics that are obviously different from other social organizations; meanwhile, the school is under the influence of other social organizations and also exerts some influence on them. From the perspective of the school itself, it has different subjects, among which differences also exist in the division of labor and duties.

The conflict of school culture can also be divided into the conflict between the school culture and the external culture, and the conflict among different subjects in the school (see Figure 6.1).

6.2.1.1 Structural elements of school culture conflicts

There are many types of classification on school culture conflicts. If we classify the conflicts based on the inside and outside aspects of the school, the school culture conflict can be divided into external conflicts and internal conflicts.

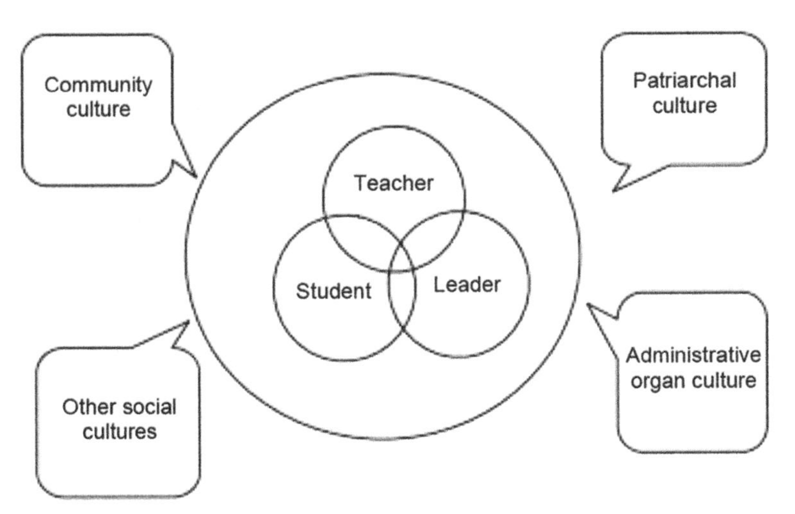

Figure 6.1 School culture conflict.

[11]Robbins, S. P., Jia Qi. Organizational Behavior [M], twelfth edition, translated by Li Yuan et al. Beijing: China Renmin University Press, 2008: 428.

The external conflicts of school culture are mainly reflected in the culture conflicts between the school and the community it belongs to, parents, the government and other social organizations. Differences between school culture and external organizations are mainly those in values, systems, ways of act and material resources distribution. Conflicts between the school and external organizations don't always exist, but are restricted by the current conditions, and present a dynamic development process along with the occurrence and settlement of conflicts.

Community is a group formed based on the restriction of geographical conditions and its member structure is relatively complex, so it's difficult to form the common values. Therefore, conflicts between school culture and community culture are not obvious in values. Most of their conflicts are reflected in systems, behavioral habit and occupation of common resources. For example, in the aspect of system, work-and-rest system in school conflicts with that of the community residents; in the aspect of behavioral habit, noises caused by outdoor teaching activities affect residents' normal life and etc.; in the aspect of common resources occupation, extension and construction of the school and etc.

Conflicts between the school and parents are caused by the students who are playing the intermediary role, which are mainly reflected in the education for students. For example, parents don't agree with the school mission, target, educational objectives, and teachers' teaching behavior, are not satisfied with the teaching quality, school environment and etc.; meanwhile, the school may disagree with parents' educational concept, educational methods and etc. Conflicts between these two parties generate because of the education of students, which is a reflection of the conflicts between school culture and patriarchal culture.

Conflicts between school culture and administrative department for education are mainly reflected in different educational programs made by these two parties and the distribution and demand of teaching resources. Both administrative department for education and the school have their own work plans. The relation between administrative departments for education at all levels and the school is like that of a board of chess and chess pieces. If contradictions occur in arrangements or the communication is not enough, conflicts will appear. On the matter of teaching resources distribution, if no corresponding teaching resources is supplied for the school development or the school doesn't make expected achievements though the administrative department for education has invested, conflict will occur between school culture and the culture of administrative department for education.

Conflicts between school culture and other organizations are mainly reflected in the differences in dominant values. For example, if the rationality, validity and education reasonability of school culture are doubted, conflicts between the school and other organizations would appear. For example, false report of the school by media, distrust and even slander of the public and etc.

Internal conflict of school culture is mainly reflected in different subcultures of teachers, students and leaders in the school.

Culture conflict between the teacher and student is mainly reflected in the teacher's teaching and the student's learning, which is specifically in the conflict of spiritual culture and behavior culture, for example, the conflict in the teacher's teaching methods and the student's accepting methods, contradiction and conflict in different ways of thinking. Gaps between the teacher and the leader are mainly in spiritual culture, system culture and behavior culture. For example, for the school's development, two parties may have different opinions; the school's rules and regulations may conflict with the teacher's behavioral habits and so on. Compared with the aforesaid two conflicts, conflicts between school culture and school leader's culture are relatively less, for the leader doesn't fully contact with each student; so, their conflicts are mainly reflected in the school's development goal and educational objectives and the student's actual development, which are always not obvious.

6.2.1.2 Cause of school culture conflict

In Lewis Coser's opinion, there are two root causes for the conflict, with one being the material cause and the other being the immaterial cause, in which the material cause is actually the objective factor that causes the conflict and the immaterial cause is the subjective factor, namely psychological factor, which causes the conflict. "Serious degree of the conflict depends on the different interactions of social structure and psychological factors".[12]

Although Coser discusses this from the perspective of the whole society, the school, as a kind of organization in the society, also conforms to this law of social development. The cause of school culture should also be divided in two parts, namely the organization structure and psychological factors of individuals in the organization.

[12]Jia Chunzeng. Foreign Sociological History [M]. Beijing: China Renmin University Press, 2008: 220.

6.2.1.2.1 *Influence of school structure on school culture conflict*

Influence factors of organization structure on the conflict mainly include the consistent degree of goals, fuzziness of division of work, resources conditions, communication situations, etc. among different organization members.[13] Correspondingly, influence factors of school structure on school culture conflict include the consistent degree of school goals and educational objectives, fuzziness of division of work among members of the school, competition for teaching resources, poor communication among teams and individuals and etc. Hereon, the influence of school structure on school culture conflict is analyzed mainly based on the inner conflict of school culture.

The degree of recognition and support for school-running philosophy, school-running objectives and educational objectives from teachers decides the school-running achievement and its motive force of development. If teachers in the school don't know clearly about the school-running philosophy, school-running objectives and educational objectives, they would just be absorbed in teaching but without knowing teaching purpose or direction. From a long-term development, this is adverse in catching up with the mainstream of era development. School teachers' recognition on the core value system of the school develops from nothing to something and then to something deepened. At first, teachers don't understand the school-running philosophy, development target and etc. and just comprehend these according to their own thoughts. In such situation, though teachers all work hard, they cannot unify to work together in harmony. When getting clearer and clearer on the school's core values, school-running objectives and educational objectives, teachers would gradually know their striving direction, but are still unable to apply such guidance into practical teaching activities though they hope to. If such situation cannot be solved in time, teachers' working enthusiasm will be bruised and circumstances such as job burnout or even resign will appear. Of course, another situation where teachers don't recognize and even oppose the school-running philosophy, development target and etc. also exists, which is very dangerous, for conflicts caused by such reason are deadly to the school since the school's development direction is involved in. Too many such conflicts may cause disintegration of the school culture or extinction of school organization.

Once two objectives of the school are established, the school principal and teachers should carry out activities centered with the objectives, so as to

---

[13] Liu Yongfang. *Managerial Psychology* [M]. Beijing: Tsinghua University Press, 2008: 311–312.

promote the realization of objectives. However, in practical work in the school, it's needed to decompose and then carry out objectives; besides, teachers have different opinions in teaching, teaching and research, management and other aspects. So, sharing out the work and cooperating with each other are especially important in the school.

The reason why division of labor is carried out is to realize the established objectives more efficiently. Division of labor is helpful for laborers to expertly grasp his/her own job skill, so as to improve work efficiency. Yet, if the post division is just carried out but not implemented by teachers, such division of labor is ineffective. Another situation is the fuzzy division of labor. Namely, everyone can do a certain job, which will cause conflicts and such conflicts are about the behavioral culture. Different people have different behavioral habits. For handling some task that has no clear division of work, different people may have different ways. These behavior differences will be reflected in culture conflicts.

Only with the clear-cut division of labor can teachers definitely carry out their work. In their process of carrying out work, teaching resources will be involved in. If teaching resources in the school are limited, competition for these resources will cause conflicts which are mainly reflected in system culture and behavior culture. For example, the competition for quota in conferring of academic titles, and the competition for activity rooms among different interest groups—one group wants to practice dancing while another group wants to rehearse drama. These conflicts are obviously caused by system arrangement and behavior differences, even though both activities are carried out to realize the development objective of the school.

In fact, contradictions mentioned above are avoidable if the school has good communication and coordination mechanism. If the occasion where the teachers don't understand the school-running objectives or educational objectives appears, the school principal should communicate with teachers in time to promote their understanding, trying best to avoid conflicts in school values; if any fuzziness occurs in the division of labor, timely remind and coordination can avoid the occurrence of behavior conflict. Meanwhile, coordinating the use of teaching resources of the school among different groups, and establishing scientific and reasonable management system can avoid the occurrence of system culture. The situation of communication and understand directly cause the elimination or aggravation of conflicts. Smooth communication will resolve contradictions, which is favorable for members to cooperate; while non-smooth communication will aggravate members' separation and the result of laying contradictions aside is to accumulate contradictions and then causes the outbreak of large-scale conflicts. The most substantive characteristics of communication are the exchange and feedback of information and emotional interaction. There are many ways to exchange and feedback information. For example, we can not only have a face-to-face talk, but also convene meetings to discuss, issue notices, set up suggestion boxes, join Internet forums and etc. to communicate.

What discussed above is the influence of objective conditions on school culture conflict, while the factor of individual psychological cognition decides whether a conflict can be a real conflict and the intensity of conflict.

6.2.1.2.2 *Influence of individual psychological cognition on school culture conflict*

Individual psychological cognition in this chapter includes social cognition and individual cognition, as well as self-regulation in this process. Social cognition is people's perception from social mentality perception of interest, demand, motivation, values and so on.[14] For human perception on social objects, one is the processing on social practice and the other is the influence on the behavior through social perception. Individual cognition refers to the individual's cognition on others, which not only includes the cognition on appearance features, but also the cognition on characters.[15] Self-regulation means the process that the original psychological states and behavioral pattern change through various ways when the individual is stimulated by the environment to adapt the external environment and keep psychological balance.[16]

There are many factors that may influence individual psychological cognition (social cognition), which can be reduced to cognition bias, scenario effect, background of the perceiver and cognitive object.[17]

Cognition bias means that the individual's certain bias affects the accuracy of cognition and makes the cognition deviate. In the process of conflict, if both parties involved in the conflict have some biases, such as stereotypes, they will generate impression on the conflict. However, if both parties can positively mediate, the conflict may be reduced; or else, the conflict may be sharpened.

---

[14] Shi Zhongzhi. Cognition Science [M]. Hefei: Press of University of Science and Technology of China, 2008: 477.

[15] Ibid, p. 484.

[16] Shi Zhongzhi. Cognition Science [M]. Hefei: Press of University of Science and Technology of China, 2008: 489.

[17] Ibid, p. 480.

Scenario effect means the influence generated on cognition after comparing with surrounding people and environment. If the contradiction is put in a relatively narrow situation, two parties of the conflict will cognize that the conflict is serious; but, if the contradiction is put in a relatively broad situation, compared with other more serious situations, the existing conflict will be relaxative.

Background of the perceiver, which includes the perceiver's original experience, values, emotion and etc., will also have effect on the cognition of conflict. The original experience is the individual's cognition pattern. If his/her original experience contains the behavior of pardoning others, the conflict will be relived if it happens; if the behavior of the other party conflicts with the original experience, the conflict will aggravate. The individual's methods of evaluating outward things are directly influenced by his/her values. If the conflict is thought to be valued and unavoidable, it will aggravate. Besides, the individual's emotional status will also affect the occurrence and development of conflicts. When the individual is good in mood, the conflict will generally be reduced; while a bad mood may aggravate the conflict. Therefore, emotion management is very important.

In addition, personal charisma, identity and roles of cognitive objects will also influence the occurrence and development of conflicts. If the object of a potential conflict is thought to be a person of strong personal charisma or experts, the perceiver may comply with his/her behavior or have new expectation, which may relieve the conflict. For example, the situation where a psychologist participates in mediating family conflict belongs to this case.

Factors that influence personal perception mainly include cognitive clue, judgment on emotion and character, and adjustment of ego imagination. The individual's cognitive clue can not only be conducted through language, but also through intonation, expression and body language. In the process of conflicts, appropriateness of these factors will directly determine the degree of conflicts; it's the same with the individual's cognition on others' character and emotion. If self-image, role and status could be adjusted timely in the process of conflicts, the settlement of conflicts will also be influenced.

Self-regulation is conducted on the basis of social cognition and individual cognition, which is a critical factor that determines the occurrence and degree of conflicts. Appropriate self-regulation is helpful to relieve and even avoid conflicts, while negative self-regulation will play an opposite effect. Through making a detailed inquiry on behavioral objective and standards, self-regulation can evaluate whether the behavior should be carried on towards the determined objective, and correct and adjust behaviors that are inconsistent with the objective. Core values of school culture should become the objective pursued by all members of the school; once any conflicts occur, take it as the objective and standard of self-regulation, which will effectively remove conflicts in school culture.

6.2.1.3 Process of school culture conflict

The occurrence of conflicts is a developing process of opposition or imbalance. Not all oppositions will cause conflicts. Only when the realistic opposition or imbalance is changed, through consciousness transformation, into some certain behavioral intention that is not always showed in external behaviors, and the behavioral intention is manifested through external behaviors, it will have some effect on the organization (see Figure 6.2).

The process of conflict occurrence can be divided into five phases, including potential opposition or imbalance, cognition and personalization, behavior intention, behavior and result.①[18] This is also applicable in school culture conflict, in which the potential opposition or imbalance is transformed into external behavior through personal cognition and then exerts different influences on the school.

Potential opposition or imbalance, also the root-cause of conflicts, in the organization is mainly influenced by organization structure and individual psychological cognition of organization members. In the aspect of organization structure, factors such as members' recognition degree on the organization's

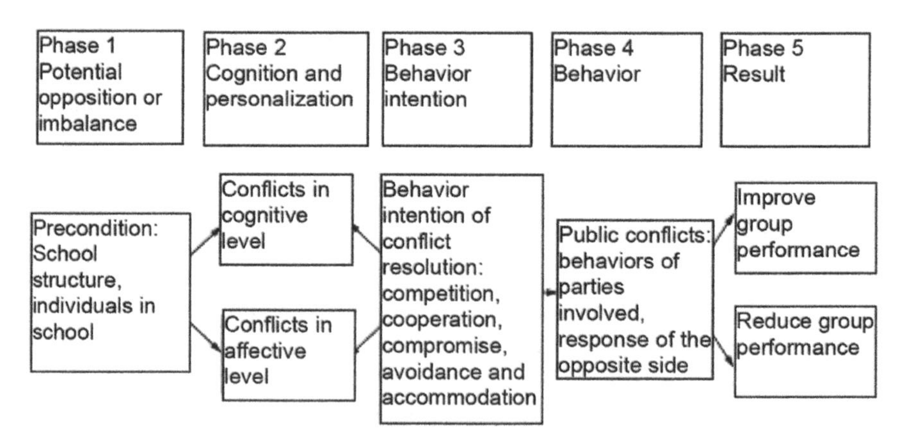

Figure 6.2 Process of school conflict occurrence.

[18]Robbins, S. P., Jia Qi. Organizational Behavior [M], twelfth edition, translated by Li Yuan et al. Beijing: China Renmin University Press, 2008: 430.

objective, insiders' division of labor, and communication channels are mainly included. In the aspect of individual psychological cognition, factors such as the personality, emotions and values of the staff are mainly covered.

When objective conditions for the occurrence of conflicts are formed, individual psychology cognition determines the degree of conflicts to some extent, for the individual or team will take the occurred potential opposition or contradiction into the consciousness range to interpret. Under the influence of objective conditions of the school organization structure, when teachers realize their interest is under the influence, they will generate the sense of conflict in consciousness. If the teacher thinks such circumstance is not as serious as it is imagined or that in intuition, or the behavior of the other party involved doesn't bring any serious consequence, the conflict will end in consciousness level, which is namely the conflict in cognition. If the teacher cognizes that such opposition in interest causes some influence on him/her and feels it's hard to accept in emotion, the conflict will upgrade to that in emotion in this circumstance.

Neither the conflict in intuition nor the conflict in emotion must transform to external behavioral expression, which doesn't mean such intuition or emotion will vanish without any reason, but will transform into a kind of potential behavior intention. Behavior intention can be divided into five types based on the consciousness on self and on the other party involved, including self-affirmation, completion desire against the other party's disagreement, cooperation desire on both parties' agreement, avoidance attitude on both parties' uncertainty, accommodation attitude on self-denial and the other party's affirmation, and compromise attitude on both parties' uncertainty.

The cognition on potential opposition is a process of repeat perceiving, in which the change of external environment will directly influence the cognition. External environment change refers to the emotion and behavior of surrounding people and the other party involved in the conflict. If the surrounding people play the role of furtherance of conflict, the conflict will generally keep aggravating; besides, the conflict behavior of both parties involved will further influence their consciousness and emotion, which will deepen the misunderstanding and confrontation in emotion to make the circumstance worse and fall into vicious circle.

External behavior of conflicts mainly include: mild difference in opinion and miscomprehension, public query or doubt, arbitrary verbal attack, threat and ultimatum, defiant physical aggression, making public adverse information of the other party, and so on.

Only when the potential conflicts transform into external behaviors will they have influence on the organization. Such influence may not only damage organization culture, but also play a positive acceleration role in the upgrading and development of the organization, with a premise that the school culture conflict is well managed.

6.2.1.4 Functions of school culture conflict

In traditional management theory, it has been engaged in avoiding the occurrence of conflict, rather than laying stress on the control of conflict; while in human relation theory, it avoids conflicts through creating favorable conditions. In fact, conflicts cannot be avoided in any organization. Conflicts within limits are beneficial to the development of the organization. Functions of conflicts are mainly reflected in the internal solidarity of groups and the balance among groups. When the objective and values in the conflict are against the values of the group, the conflict is dangerous and hard to mediate, and has no positive functions.①[19] Such conflict belongs to the negative one.

Positive conflicts can define the limit among groups and enhance cohesive force inside the group; and also play the role of safety valve to allow members of the group to vent their unhealthy emotions through conflicts. Meanwhile, positive conflicts also play an important role in maintaining the stability and balance of the whole society.

6.2.1.4.1 *On the cohesion function of culture conflict on school*

Georg Simmel thinks conflicts are beneficial for the cohesion in groups; detestation and mutual opposition are able to maintain the whole system through establishing balances in constituent parts. Such cohesion function is mainly reflected in maintaining limits of groups, promoting internal solidarity of organizations and improving diversity of organizations.

School culture is helpful to maintaining the limits. Differences and conflicts between the school and other organizations or groups cause the differences in values and behavior standards between the school and other organizations. Just these conflicts and differences, to some extent, definite the limits between the school and external organizations and make the school become a unique organization with its own characteristics and features. In the developing process of culture conflicts between the school and external

[19]Coser. Functions of Social Conflicts [M]. Translated by Sun Liping. Beijing: Huaxia Publishing House, 1989: 67.

organizations, common value priorities and recognitions come together, which contributes to the relative stability of school.

School culture conflicts are helpful to enhance internal solidarity in school. The conflicts between one group and other groups are able to activate the vigor in members, so as to enhance internal solidarity of groups. This is of the same principle with the situation in which when a country is invaded by foreign enemies, different political parties will temporarily abandon prejudices and band together to resist the enemy. Of course, such external conflicts don't always occur, but the groups must realize the existence of threat. For example, if students in the school haven't gotten satisfactory results in one test, source of students enrollment will be threatened; in such circumstance, conflicts occur in this school and other schools; these conflicts and threats will make teachers of the school unite as one to work hard to improve the school's reputation to promote the school to develop better.

School culture conflicts are helpful to enrich the diversity in school culture and promote solidarity in a wider range. In spite there are common objective and value basis in school, it doesn't mean that contradictions and conflicts don't exist; while frequent conflicts in school don't mean they are not united. On the contrary, these conflicts indicate that a high-level debugging is going on among members of the school. To some extent, differences and conflicts among members of the school reflect the diversity in school culture, and are also a reflection of humanization in the school. If these conflicts and differences can be expressed and the expression through culture conflicts can enhance the cohesion and comprehension among members, it will be favorable for establishing the solidarity in a wider range and improving cohesion among members.

6.2.1.4.2 *On safety valve function of school culture conflicts*

Coser thinks that conflicts can play the role of safety valve in a group. Safety valve is the attack against replaced objects and the hostile energy release in other types of activities.[20]

It's impossible for individuals in any groups to express his/her own hostile thoughts or emotions. However, the accumulation of negative emotions may cause conflicts. If these negative emotions can be released instead of accumulating limitlessly, conflicts can be avoided to some extent. In the school, the permission of expressing culture conflicts is helpful to avoid

[20]Coser. Functions of Social Conflicts [M]. Translated by Sun Liping. Beijing Huaxia Publishing House, 1989: 26.

more serious cultural divide (behavior of value subversion). The expression of opposing emotions or conflicts among school members is of important significance, which plays the role of safety valve for the school.

There are many ways to give vent to conflicts. The individual can vent conflicts not only on the original hostile objects, but also on replaced objects. There are three types in expressing negative emotions: first, express hostility directly on the individual or group that cause setbacks; second, throw the hostility behavior to the replaced object, for example, point at one but abuse another; third, provide activities to gain some satisfaction and release the tension state, in which no objects or replaced objects are needed and an outlet is provided to release the hostility; for example, the method of physical exercise can play a part in this.

In fact, the above three types can be regarded as two types. First, vent conflicts directly on the setback source, which is helpful to vent the discontent thoroughly but will do some harm to the group. Second, find replacement in venting conflicts, in which the individual can choose method replacement or objective replacement. Method replacement refers to adopting other methods to vent the discontent but not causing conflicts; while objective replacement refers to looking for other objectives to vent the discontent on, which is generally disadvantageous for a thorough problem solving, for the hostility on the setback source isn't removed fundamentally. Besides, safety valve system impedes the social relation change which is to adapt to the changing environment. Hence, the satisfaction degree it provided to the individual is just part and temporary.①

If culture conflicts occur among different members in school, the school should allow the conflict expression instead of suppressing, in which different solutions should be chosen in accordance with specific environmental conditions and characteristics of two parties involved in conflicts. For direct vent on setback source, problems can be solved timely if the communication is smooth, while more server conflicts will break out if there are barriers in communication. Indirect emotion vent is able to moderate conflicts temporarily, but cannot solve problems fundamentally.

The adoption of safety valve system is determined by the situation that whether hostility expression is allowed to express in a group. If the hostility expression is not allowed to express, safety valve system is needed for members of the group; if the group is democratic and members can express their emotions freely, safety valve system is not that needed.

Through the above analysis, it's showed that the premise of conflicts function lies in timely control and adjustment of conflicts. Only when the conflicts caused are not aiming at the objective, values and etc. of the organization can they be positive conflicts. In such circumstance, the premise of realizing the function of conflicts lies in the inevitability to admit conflicts, so as to commonly solve problems and promote the development of the organization. Therefore, solidarity is established not only on the base of homogeneous values, but also on the base of differences caused due to social division of labor.

6.2.2 Solidarity of School Culture

In school cultural mechanism of action, there's another indispensable mechanism in addition to the conflict – solidarity. Solidarity is a final state that needs to be achieved in a circular process and also a state in which both parties cooperate and connect with each other again after the resolution of conflict.

6.2.2.1 The type of solidarity of school culture

Solidarity is a connected state, which is first put forward by Emile Durkheim, who used social solidarity when discussing the function and purpose of social division of labor. Social solidarity not only refers to the direct face-to-face social interactions among people in daily life, building friendship, marriage and other fixed relations, or the psychological states and behave patterns of caring each other and helping each other in interactions, but also includes a wider range of phenomena. The basis of social solidarity is the common values and common moral standards of social members, and base on the strength of the awareness of such common values and common moral standards, social solidarity can be divided into mechanical solidarity and organic solidarity.

Solidarity of school culture refers to the connection mode among school members, which not only includes direct face-to-face interactions, but also psychological states and behave patterns, and even value orientation, etc.

Social solidarity is divided into mechanical solidarity and organic solidarity, but cultural solidarity is doomed not to be the mechanical solidarity, but the further development of organic solidarity, namely a new solidarity mode formed with the arrival of an era that people develop themselves comprehensively and pursue life-long education by taking the cultural element of value orientation as the connection.

The change from mechanical solidarity to organic solidarity is a historical development process and also a social development process in which mechanical solidarity keeps retreating, and organic solidarity keeps replacing mechanical solidarity.

The underlying factor that determines the transition from mechanical solidarity to organic solidarity is the social division of labor, which directly leads to changes in the power balance between collective consciousness and personal consciousness (as shown in Figure 6.3).

6.2.2.1.1 *On mechanical solidarity*

In the period when the social division of labor is not distinct and personal consciousness is weak, differences between people are small and collective consciousness almost completely controls personal consciousness; in such situation, collective consciousness and values become the bond that connects people together, which is mechanical solidarity.

Mechanical solidarity is a connected state between individual and individual, individual and group, and group and group build on the basis of common emotions, morality, faith or values characterized by unity or attraction, which combines homogeneous individuals together through strong collective consciousness.[21]

Mechanical solidarity is a primitive connection, which requires individuals to directly blend in the group without personal consciousness, in which personal opinions and consciousness form the common values of the organization or group. Therefore, the ideology and values with personal characteristics have no space to live, and such connection goes against the development of personality and individual character development of humans. School

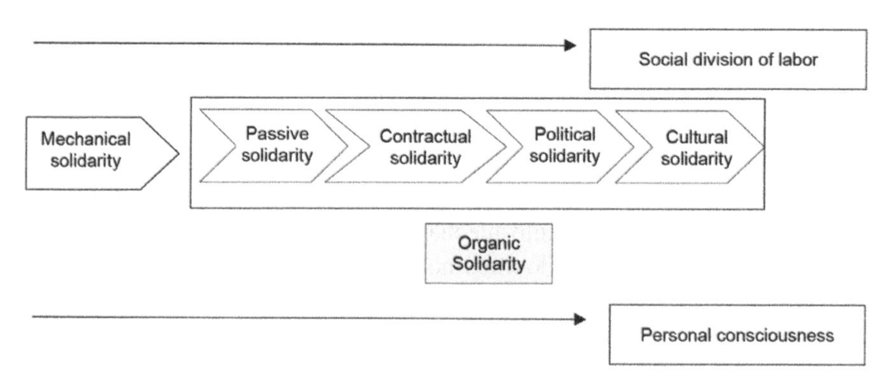

Figure 6.3 Development stages of social solidarity.

[21] Jia Chunzeng, The History of Foreign Sociology [M]. Beijing: Renmin University of China Press, 2008: 116–117.

members are not completely homogeneous. Although they need to have an identical value pursuit and a common goal, they may have other pursuits in addition to work for their school life is just one part of their entire life. Thus, school members are impossible to be completely classified into the school, because they still have their families, their small groups and their social circles. Therefore, the solidarity in school is impossible to be mechanical solidarity.

As the social division of labor arises, the social connection with the bonds of occupation is gradually established and different occupation groups have different values. At the same time, with the awareness of personal consciousness, the territory of common values in the entire society starts to reduce, but certainly not disappear. It's just that the power balance between personal awareness and social common values has changed, and collective consciousness cannot overwhelm personal consciousness anymore; in such situation, people need each other due to the social division of labor. Thus, the organic solidarity comes into being.

6.2.2.1.2 *On organic solidarity*

Organic solidarity is a connection built on the basis of the social division of labor, in which social members are concordant and interdependent.[22] In the same society, people doing the same labor have such connection.

Organic solidarity is the result of social division of labor and is built on the basis of personal difference, which can show the personality and characteristics of members. Organization members only partially belong to the organization, but beyond that, organization members still have their own sphere of activity and can realize self-development. "Each person has a sphere of action which is peculiar to him, the ability to reach his own realm, and his own personality."[23] The solidarity in school belongs to organic solidarity. School's requirements and constraints on its members are based on the job duties and performance, while school members also connect to the school due to their job (the social division of labor). Beyond that, school and its members are independent.

The basic reason for the birth of organic solidarity is the differences caused by the social division of labor. In different stages of the development of social division of labor, organic solidarity also has different forms.

---

[22]Same with ①, P118.

[23]Emile Durkheim, The Division of Labor in Society [M]. Translated by Qu Dong. Shanghai: SDX Publishing House, 2000: 90–91.

6.2.2.1.3 *On passive solidarity*

Private property becomes an intermediary in the form of object to intervene in people's social relationships. The form of object fixes the relationships between people to ensure the rules of people's behavior, and this constitutes passive solidarity. Such solidarity determines the connection between people by defining the relationship between different people and a certain object. Obviously, such solidarity form is improper in school culture.

6.2.2.1.4 *On contractual solidarity*

With the professionalization of division of labor, different labor groups in society are gradually divided, and each group has its own values and pursuits, i.e. professional ethics, but any single social group can't exist independently, so differences and cooperation are both needed. In order to ensure a balance in interests, each group needs to give sincerity in the process of cooperation and bear certain obligations. In such situation, the establishment of contract solves such conflicts effectively, so people enter into the era of contractual solidarity. Contract is the result of the game between stakeholders. In school, the establishment of various rules and regulations is such a result of game, which reflects the responsibilities and rights of different members in the school. Certainly, such culture based on regulations can regularize school members, make them clear about their responsibilities, and better promote the harmonious development of the school.

6.2.2.1.5 *On political solidarity*

In the process of making various contracts, a certain value standard still needs to be ensured, and an organization is also needed to restrict and manage both parties of the contract, so various political organizations and special professions are established. Such political organizations are built based on the common values of the entire society, with the purpose of maintaining social stability, making and enforcing laws for the society. School is not a political organization and can't bear the responsibility of lawmaking, so the solidarity in school is impossible to be political solidarity.

Various rules and regulations established under contractual solidarity and political solidarity is able to restrict society or organization members to a certain extent, but with the speedup of information transfer and the gradual establishment of life-long education system in society, the range of labor force flow is gradually increasing and speeding up. In such case, maintaining the stability of a group and attracting more high-level talents to join the organization becomes an important challenge for the survival and development of the organization. In order to better promote the development of the organization, it is necessary to build a new connection mode among members, and that is cultural solidarity.

6.2.2.1.6 *On cultural solidarity*

Cultural solidarity refers to the connection built with the bond of common values, ideals and faith etc. among members. The consistency of values emphasized in such connection is the consistency after clash and collision, but not the same with that in mechanic solidarity. The territory built by cultural solidarity is limited in working places, and beyond work, any organization member has his or her own lifestyle, so cultural solidarity belongs to organic solidarity. Certainly, the connection bond in cultural solidarity is the common values, ideals or faith of the group members, which are the intermediary of thoughts and consciousness rather than the object, so such connection is not passive solidarity.

The common values, ideals, or faith etc. that connects school members are finally reflected in different system cultures, and clarifies the labor division and duty of members through the coordination of system. Such emphasis on system has used contractual solidarity and its form for reference, and leads and improves to a higher level of value on its form.

The ultimate purpose of school culture is to realize contractual solidarity and cultural solidarity. Solidarity of school culture is the system culture in the school formed by carrying out the core value pursuit in actual education and teaching led by the common core values approved by all school members. School system culture doesn't exist in isolation, but has some relation with various teaching activities of school members. So, with the provisions of system culture, core value of the school and various activities in the school can be connected, which contributes to the formation of school behavior culture. In addition to the system and behavior that can express the core value of school, the material environment in school can also express the core philosophy of school and also becomes a part of the school culture. When all of such coordination consistently centers on the common core value of school, the solidarity of school culture realizes. The solidarity form with the bond of common cultural identity is necessarily stable and firm, and also conforms to the law of personality development of humans. It can be deemed that cultural solidarity is the highest level of school development.

6.2.2.2 Effect factors of cultural solidarity

The basic reason why people come together and build connection within the group and between groups is that they have common social consciousness and values. However, with the progress of productivity, the division of labor comes into being in human society, and the birth of division of labor facilitates the division of common social consciousness. In such situation, personal consciousness is acknowledged, which greatly promotes the change and development of solidarity types. So, the social division of labor is also an important reason for the birth of social solidarity.

The birth and emergence of solidarity of school culture, on the one hand, is due to their common social division of labor, as they are all engaged in school teaching as a teacher, and on the other hand, is due to the common value they share with the school. The characteristics of social division of labor determine their job as a teacher, while the recognition of school's values determines the school they work for.

6.2.2.2.1 *Impact of common consciousness and values on solidarity of school culture*

The power balance between personal consciousness and values and the common social consciousness and values of the whole society determines the type of social solidarity. When it is mechanical solidarity, personal values are consistent with collective values, and personal consciousness, if it exists, is oppressed by collective consciousness. When the power balance between personal consciousness and collective consciousness changes and personal consciousness gradually overcomes collective consciousness, differences between people will emerge; individuals will have different pursuits and choose different professions and work organization, which brings humans into the era of organic solidarity.

So is the establishment of solidarity of school culture. Different individuals in society have different selection of values. They can select to be engaged in education, medical or legal work. Once someone selects to be engaged in education, it indicates that at least he doesn't repel to education; thus, all the people engaged in education will have the common value pursuits; or it can be deemed that professional ethics and pursuits generate from such "common value pursuits".

Through such most basic value recognition, people build solidarity between each other, and such solidarity is the recognition of professional values. However, every school has its uniqueness, with different educational philosophies and expressive ways of objectives. So, the recognition of school-running philosophy and objective of a school is necessary to gather members in the field of education in this school. Based on the recognition of school-running philosophy and objective of a school, members in the school build a stable connection with each other, which is an important effect factor for the solidarity of school culture.

6.2.2.2.2 *Impact of the social division of labor on school culture*

The social division of labor also impacts the establishment of solidarity of school culture. The normal operation of a school not only needs the teaching of teachers, but also needs the management and leadership of leaders, as well as the guarantee of educational administration of educational administrators; and different divisions of labor also exist in teachers' work, which is reflected not only in the differences of grades, but also in the differences of subjects. The different divisions of labor of different personnel in school reflect the differences between members, and it is such differences that determine a certain person or a certain kind of person cannot complete all the tasks in school and they have to cooperate with each other to complete them.

With the establishment of cultural solidarity, the school can operate more effectively so as to facilitate the enhancement of school cohesion. Thus, more widespread and indeterminate individuals can be included in a group, and the concrete behaviors of people can be concordant with each other.[24] However, such solidarity still contains the hazard of collapse.

6.2.2.3 Collapse of solidarity

Social solidarity built on the basis of the social division of labor and common values is an ideal status. In real life, various limits usually lead to incompatibility in social solidarity. When the division of labor cannot contribute to solidarity, the relationship between various organizations falls into anomie[25] before defined, i.e. collapse of solidarity. The main reasons for the collapse of solidarity are competitive plan, extreme individualism, abnormal social division of labor and etc.

School is one of the social organizations. So, the reason for the collapse of solidarity of school culture can be summarized in the intensification of school competition, individualism in school and abnormal social division of labor.

---

[24] Emile Durkheim, The Division of Labor in Society [M]. Translated by Qu Dong. Shanghai: SDX Publishing House, 2000: 68.

[25] Same with ①, P328.

6.2.2.3.1 *On the intensification of competition*

With the increase of population and a gradual intensification of competition in society, the conflicts between members are also strengthened, which will damage social solidarity. It is the same in school. That school members pay too much attention to competition will lead to interpersonal strain, and such intense emotion goes against the solidarity and cooperation between members. Especially if vicious competition appears, the consequences will be horrific.

In order to solve such conflicts, it will be necessary to professionalize and further subdivide the labor in society to reduce the competition for survival between people and meanwhile enhance the inter-dependent and cooperative relationship between professionals. In the school, differences among teaching posts, management posts and educational administration posts are distinctive with detailed division, so they won't have much competition with each other. However, people in the same post will have relatively much competition with each other. So, it will be conducive to the establishment and consolidation of school solidarity to avoid ultra competition in the same post, emphasize on the competition based on cooperation, jointly promote the development of school through exploiting different potentials of each members, and guide different school members to cooperate by taking advantage of each other, which is not only a new understanding of the subdividing of labor in school but also a kind of cooperation built on the basis of differences.

6.2.2.3.2 *On extreme individualism*

With the enhancement of personal consciousness, the binding of collective consciousness will be relatively reduced. In the whole, if little extreme individualism emerges, such individualism may help the individual obtain certain achievements in a short period as it prioritizes personal short-term benefits; but in the long run and in terms of the collective interests, it goes against the establishment of social solidarity. This is also true in school. Irreplaceability of members may lead to the expansion of personal consciousness of individual members, which goes against the cooperation and the establishment of solidarity.

6.2.2.3.3 *On abnormal social division of labor*

Abnormal social division of labor will damage social solidarity. Emile Durkheim thinks that abnormal social division of labor is mainly divided into three forms: the first is dividing labor through subjective impression and judgment without learning the actual situation, which lead to the new social division of labor being unable to adapt to the reality situation and out of touch with the reality; the second is forcing to divide labor without following the standard of personal skills and knowledge but through controlling other social resources; and the last is emphasizing too much in the division of labor and ignoring personal feelings so that personal activity is unable to be maximized. These are all inimical to social solidarity.

So is the division of labor in school culture. The presentation of any new philosophy and the issue of new rules shall take the real situation into consideration, but not be estimated by subjective forecast, or else the ideal may be disconnected with the reality. Before the old member relationships are not completely weakened, the new built relationships are not likely to be adapted. For the establishment of various relationships in school culture, it shall be based on different capabilities, provide fair and flexible environment, and establish platforms for the development of school members, taking personal interest of different members into account to give everyone's capabilities to play.

Such a series of anti-solidarity conditions, to a certain extent, are the conflicts we've mentioned above. In the three reasons leading to the collapse of social solidarity, both the intensification of competition and abnormal social division of labor are caused by the alienation in organization structure, while the extreme individualism can be attributed to the factor of personal psychology. This is the same with the reason of conflicts. Therefore, we can believe that it will be a perpetually circular, ever-changing and developing process from the emergence of conflicts to the realization of social solidarity as well as the emergence of conflicts in social solidarity.

6.2.3 Conflict-Solidarity Spiral

Talcott Parsons divided functions of the social system into the following several different processes, including orientation mode, types of action, cultural interactions of actors with different orientations, interactive institutionalization, and status, roles and standards (see Figure 6.4).

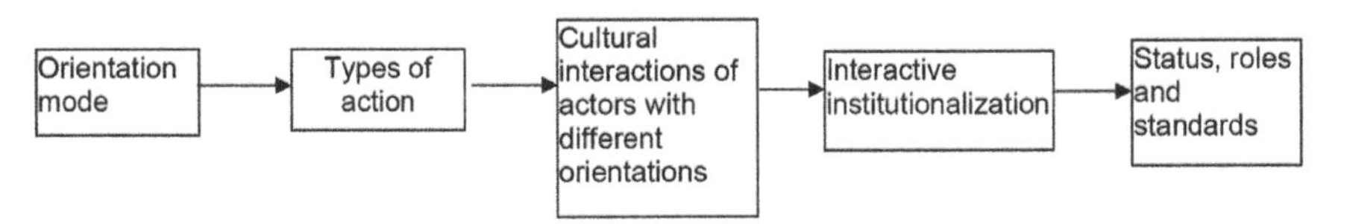

Figure 6.4 Social system theory put forward by Parsons.

In the social system theory put forward by Parsons, orientation mode refers to the behavior orientation chosen by actors under the influences of motivations, values and situational conditions. There are three types of motivations, which are the cognitive type that pursues information, emotional type that pursues emotion and the assessment type that pursues assessment results. Values can also be divided into three types, including cognitive type that takes objectivity as the value standard, appreciation type that takes the aesthetics value as standards and moral type that takes absolute correctness and errors as the value standard. Actors perform different types of action in accordance with a certain kind of the strongest motivation and values and a certain kind of combination mode. Since motivations and values of each person are different, types of action produced are also different. Interacting among actors of different orientations gradually institutionalize these interactions, which contribute to the stability of different individuals' status and roles, as well as the establishment of certain standards.

So are the processes of production, development and stabilization of school culture. Any school culture is set up by school members under the guidance of different motivations and values and meanwhile is limited by objective factors in surroundings. Different motivations and value orientations among school members can form uniform school culture only through interactions and conflicts and the school culture will be continuously carried out in all kinds of systems of the school. As a result, the institutionalized school culture will exercises its function to define and stipulate the core values, work distribution and responsibilities for each member of the school; meanwhile, it explicitly stipulates roles and status of school members through all kinds of physical forms and various activities in the school.

From the perspective of social system theory put forward by Parsons, it can be found that the occurrence mechanism and mechanism of action of school culture, can be divided into two parts, namely the process stepping from conflict to solidarity (see Figure 6.5).

The conflict stage of school culture includes orientation mode and cultural type. This is consistent with the conflicts presented by Coser. In Coser's opinion, social structure and personal psychological factors leads to the production of conflicts and these two causes leading to the conflict are essentially subjective and objective conditions. The factor of social structure belongs to objective conditions, namely the situational factors stated by Parsons; while personal psychological factors belong to subjective factors, namely individuals' different motivations and values emphasized by Parsons.

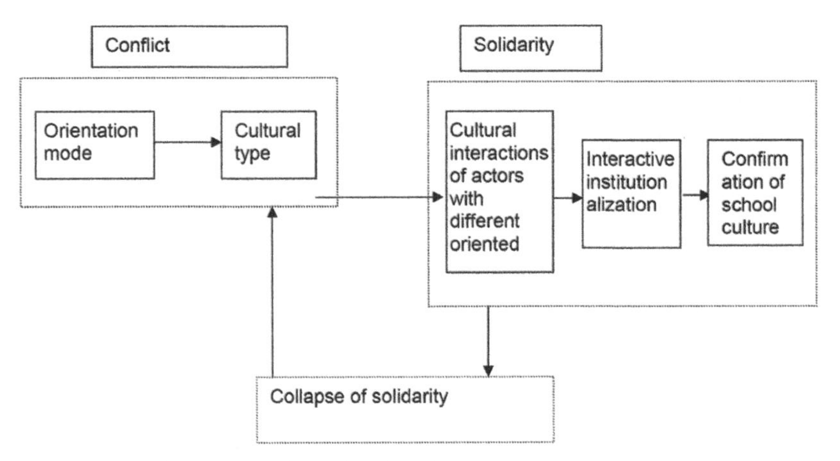

Figure 6.5 Mechanism of action of school culture conflict-solidarity.

The solidarity stage of school culture includes cultural interactions of actors with different orientations, interactive institutionalization and confirmation of school culture. Emile Durkheim pointed out that different social divisions of labor led to the occurrence of organic solidarity in the society. The social division of labor is also a kind of gradual processes of selecting and differentiating, and actors' interactions stated by Parsons are also a pattern of manifestation of social divisions of labor. The occurrence mechanism and mechanism of action of school culture are processes of mutual conflict among different divisions of labor and culture orientations among school members. Motivations and values of different members become institutionalized gradually and form stable social status and confirmed roles, which is also a process of forming an inherent norm, differentiation of vocations in the society, and confirmation of occupations, as well as the forming process of solidarity of school culture.

The mechanism of action of school culture is a process stepping from conflicts towards solidarity. However, with the limitation of a variety of conditions, there are also some factors which will lead the solidarity stepping towards collapse. Emile Durkheim deems that, among the factors which lead the solidarity stepping towards collapse, the weakening of collective consciousness and the mightiness of personal consciousness, and even abnormal social division of labor, all can be summarized as changes of social structure and personal psychological and these are exactly the source of producing conflicts. As a result, to some extent, it can be deemed that the collapse of social solidarity is also the source of social conflicts.

The occurrence mechanism of school culture is a spiral ascending and continuous developing process, in which conflicts occur along with changes in internal and outside conditions and gradually converse into school solidarity to promote the development of schools. But continuous changes of outside environments and conditions urge the spiral mechanism of school culture to step from conflict towards solidarity to advance and develop continuously; in this process of continuous conflict and solidarity, the school culture improvement realizes its own purpose: solidarity of school culture and culture prosperity.

6.2.4 School Culture Conflict Management

Solutions of school culture conflicts can be mainly divided into two kinds: one is the conventional conflict management and the other one is the emergency conflict management.

6.2.4.1 Conventional management of school culture conflicts

Conventional management of school culture conflicts refers to the normalized, regularized and standardized management realized by the school for managing various conflicts occurred in daily life through establishing effective systems and procedures to give play to the synergistic effect of various parties related in the school under normal state. Through the conventional management, the school can early warn about and guard against some problems and contradictions and solve them as soon as the conflicts occur by timely motivating all elements of power. The daily management of school culture conflicts is mainly manifested on setting up a common vision, forming a model of joint participation and setting up a smooth communication mechanism.

What the conventional treatment of conflicts emphases is to solve conflicts through the comprehensiveness and predictability of daily work. Therefore, the conventional treatment of conflicts in school culture construction means that all members should form a relatively uniform vision and set up a set of comprehensive and specific system.

Set up a common vision. In school culture construction, the establishment of a common vision can provide striving directions for school members and also the lasting power for the development of the school. Core values of a school are the common vision of the school. Create a kind of cultural atmosphere by virtues of factors such as buildings, banners, slogans, legendary stories, ceremonies, and declarations of principle to infect school members through this kind of atmosphere to solve conflicts existing in thoughts.

The model of joint participation: under many circumstances, when school members can jointly participate in the school culture construction, many conflicts and contradictions can be solved in the process of participation. In the process of participation, different conflicts and contradictions will emerge; thus, by expressing and discussing suggestions, conflicts can be solved.

Communication mechanism: a good communication mechanism not only can ensure smooth communication of information, but also can timely express different suggestions and opinions at any time.

6.2.4.2 Emergency management of school culture conflicts

The emergency management of school culture means that during abnormal period and under a circumstance that conflicts have already emerged obviously or an emergency circumstance occurs, the school control the extent and the scope of conflicts timely and effectively through a variety of irregular management measures and means to limit the intensification and upgrading of conflicts and furthest reduce damages to and recover normal social order as soon as possible. The characteristic of the emergency management is finding similarities and features of conflicts quickly and accurately to solve problems effectively under complex circumstances full of changes.

When there is a conflict occurring in the school, the first action is to diagnose the conflict to tell where the source of the conflict comes from, which scope the conflict belongs to, and members who are involved in the conflict, as well as to predict the best and worst results of the conflict and consider whether it can use other experience for reference etc. to motivate all active factors to solve conflicts as far as possible. Determine solving strategies for conflicts on this basis.

The solving strategies of conflicts generally include forms such as expanding resources, establishing a system for appealing, changing models of interactions, altering the reward system, merging, role classification, mediation of the third party and so on. When there is a conflict occurring in schools, different solving strategies can be chosen in accordance with concrete circumstances. For example, if the quantity of teachers and students of schools is more than the carrying capacity existing in schools, expanding resources will be an appropriate method for solving the problem; if teachers are widely dissatisfied with the appraisal mechanism, reforms of the reward system will be imperative.

In the process of the emergency management of school conflicts, middle-level cadres play an important role. They generally play following roles in the management of conflicts: discoverers of conflicts, punch bags of conflicts and mediators of relieving conflicts. Middle-level cadres directly come into contact with the daily work in schools and they can discover conflicts earlier if any problems occur; thus, contradictions can be timely found by virtue of their sensitivity and solutions can be put forward as far as possible. Of course, once there is an intensive conflict occurring, they are also objects directly complained by teachers and students, which requires that they should have enough patience to tolerate and understand. Then, they should mediate and relieve those contradictions.

Both from the perspectives of theories and practices of the school culture improvement, the target of school culture improvement is to achieve the solidarity and prosperity of school culture to promote humans' development and the progress of organizations. The process of realizing solidarity of school culture is a process stepping from conflict to solidarity. In the process of realizing solidarity and prosperity of school culture, it is necessary to guarantee the reasonability and legality of school culture; and in the concrete practice of the school culture improvement, it is necessary to set up a complete conventional management mechanism and a complete emergency management mechanism of cultural conflicts. Only by this, there can be more vitality in school culture and the development of schools can be provided with ambitious directions and inexhaustible power.

6.3 School Culture Construction

The school culture improvement and construction process is to achieve cultural unity and prosperity, heading for an ideal condition on the basis of original culture. Every school thirsts for possessing a strong culture. See the interpretation of spectral theory of school culture for details.

The school cultural prosperity means the progress and improvement of school culture on the basis of original culture and such progress requires continuous thinking and efforts. The school cultural management and construction should be rational and legal, namely reflect the rationality of value and instruments in the aspect of rationality, and meet the requirements of laws and regulations, and guide and bind the school members in the aspect of legality.

6.3.1 Construction of Value Rationality

Firstly, the school culture construction should have clear values. There are many kinds of values in human society, but each school should possess only one reasonable and legal core value.

In general, there are two ways for school to determine its values. The first is to choose a kind of reasonable value pursuit as the core values, and conduct and organize all activities based on the core values. The second is to summarize the existing actions, activities, even the system, etc. to determine a core value pursuit conforming to the regulations of the society. The first is in deduction form, and the second is in summarization form.

It is required to take the mission of the school and its characteristics into account when determining the core value in the form of deduction, and choose the core value in four aspects as follows: first, it's education on personality, which includes respecting, cultivating, developing the personality of students, self-realization etc.; second, it's cultivation on sociality, which includes vocational ability, cultural character, political identity, responsibility of family life, etc.; third, it's national identity, which includes ethnic identity, national identity, history identity, etc.; fourth, it's awakening of the human consciousness and the formation of such common values as independence, freedom, equality, tolerance, peace and sustainable development.[26] It is understandable to choose the value from any aspects, but the most important is to provide a development direction and be the power to lead the school development. However, in the process of choosing the core value for the school, it is required to pay attention to improving the discussion and research on the school culture, be adept in learning the culture essence at all times and in all countries as well as that in school historical tradition, and reflect the characteristics of times and nation.[27]

It is required to summarize the specific rules and regulations, teachers and students' activities, school cultural landscape, etc. when determining the core values in the form of summarization. There may be several value requirements or regulations in the results of summarization. For example, focusing on details, namely accuracy, analyzing and focusing degree on details; result-oriented, namely paying attention to the results but not process; employee-oriented, namely the influence degree on organization members; team-oriented, namely organizing work by the team not individuals; aggressiveness, namely being aggressive and competitive; stability, namely maintaining the status quo; innovation and risk tolerance, namely making innovations, bearing risks, etc.[28] The identity on one or several value pursuits

[26]Shi Zhongying, Core of School Culture: Value Construction [J], Educational Science Research, 2005 (8): 4–7.

[27]Ditto.

[28]Stephen P. Robbins, et al, Management Theory [M], Trans. Sun Jianmin, et al, Beijing: China Renmin University Press, 2009: 58.

among them provides a direction and value guidance for the development of the organization to some extent. However, in this process, it should be noted that there shall be no conflict among different value pursuits, and all these value pursuits shall form a logic, orderly and organic value system. See Chapter 1 for details.

The determination of the core values of the school means the determination of the direction of its development and cultural management. What's more, it is also crucial to decompose the core values, enrich its connotation to show the characteristics of the school, and finally put the core values into the actual education practices. Except for core values, the school culture includes system culture, behavioral culture and material culture which should reflect the core values of the school admittedly as well as their characteristics in various activities. The strong and wonderful school culture is sure to be not only the mere form for the publicity of the school but also the source to provide continuous power for the school to develop.

6.3.2 Instrumental Rationality Construction

Generally, the instrumental rationality of school culture is reflected in efficiency problems occurring in the process of carrying out core values in specific systems, activities, behaviors, and other aspects. The corporate culture construction can be used for reference to better reflect the effect of school culture on efficiency.

In the research on some excellent corporates in America, Thomas Peters and Robert Waterman found that the success and culture are connected closely. In other words, their good culture brings success. Such cultures reflect the characteristics of system culture and behavioral culture to a certain extent, and worth learning to prosper the school culture.

The excellent cultural characteristics of successful corporates include: (1) Focusing on solid work. Correspondingly, this is to carry out the objective management in school and continuously inspect the realization degree of objective, but not to lay aside and neglect the determined objective. (2) Getting close to customers. Correspondingly, this is to concern about students, consider the comprehensive development of students, and strengthen the cooperation with parents. (3) Improving the productivity of labor by people. It is obvious in school, because the educational activities in school are the communication and influence activities among people. It is inappropriate to establish too many tough orders, but it is required to believe in teachers and allow them to express their thoughts. It is also very important to concern about the welfare of teachers. (4) Value-driven efforts. The value-driven school development is a major requirement of school cultural management and construction, so the way to form a value-driven behavior model requires school members to not only pay attention to the value pursuit of the school, but also fulfilling such value through substantial efforts. (5) Focusing on own business. The school is given so many functions by the modern society, but the school should understand that its important mission is to teach students and promote a better development of students. (6) Simple form. The simple here means streamlining the workforce. However, it is to streamline the teachers, management team, and logistics team in school. (7) Loose-tight characteristics. The proper school culture characteristic refers to a tight common value pursuit and a loose professional autonomy.

The value and objective rationality of school culture must be unified, in other words, the inner value pursuit of school culture should be unified. It is not an easy job to unify the core value, behavioral culture, material culture, and system culture of the school. Under the ordinary circumstances, college experts have an advantage over such aspects as the height of foothold, the extent of examination and depth of analysis, with a stronger capacity in refining core thoughts, interpreting conceptual framework and setting up a general framework; while the participators of middle and primary schools have an advantage over such aspects as perceiving the actual scene, adjudging the realistic relationship and understanding the actual results, with more targeted capacity in choosing specific ways, designing operating modes and handling actual contradictions.[29] Therefore, the school can draw support from outside the school including such organizations as universities, research institutes, administrative authorities and other social groups to promote the rationality of school cultural management.

6.3.3 Legality Construction

The functions of school culture are to arouse the enthusiasm of teachers and students, clear the development objective of school, and finally promote the development of school through the power of culture. How does the school culture develop its demonstration effect and promotion action?

Part of cultural unity can be realized through contract unity, because the influence of culture can bind members through the visible rules and

[29]Wu Kangning, From Interests Combination to Cultural Integration: Deep Collaboration with University and Middle and Primary School [J], Journal of Nanjing Normal University: Social Science Edition, 2012 (3): 5–11.

regulations, and exert influence on members by such various subtle means as symbols and ceremonies. The establishment of rules and regulations can take effect instantly to a certain extent, but may not be completely accepted by teachers and students; especially, when these systems are not established by teachers, the effect will be worse. However, there is an exception, involving the participation of new members. Therefore, the identity level of school culture should be a threshold for the admission of new members and it is essential to spread the culture of the organization to new employees through selection standards. After entering the school, new employees are required to consider integrating into the school culture by other means as soon as possible.

The school culture unity is more likely to be realized through the influence of invisible cultural objects on school members, including various ceremonies, events, and stories of excellent teachers and students, special communication style of the school, leadership style, etc. For instance, the "coming-of-age ceremony" for students and graduation ceremony are ceremonies for students to express "gratitude" to teachers and parents; the school's presenting a bunch of flowers to teachers on Teacher's Day can not only reflect the school's qualities and but also provide spiritual impetus for the development of school members. Learning from the excellent teachers and students not only is suitable to their actual circumstances, but also provides them an achievable target. Moreover, the special discourse system and expression form of the school can increase the organizational identity of teachers and students, reflecting the uniqueness of the school. The work style and expression model of the leadership have a profound and lasting influence on the school. In almost all schools, the work style and expression model of teachers are the same as that of the principal of the school to a great degree.

If the culture of the school is conforming to value standards and legal regulations, and has a subtle influence on the teachers and students of school, it must be a high-quality culture with continuous features.

References

Monograph

[1] Chen Yukun, *Educational Evaluation* [M], Beijing: People's Education Press, 2001.
[2] Chen Zhenming, *Analysis of Public Policy* [M], Beijing: China Renmin University Press, 2003.
[3] Gomez, et al, *Management Theory – Human · Performance · Transformation* [M], Trans. Zhan Zhengmao, et al, Beijing: Posts & Telecom Press, 2009.
[4] Jin Di & Wang Gang, *Education Assessment and Measurement* [m], Beijing: Science and Education Press, 2001.
[5] Li Guolin, *Taiwan School Culture in Social Transformation* [M], Fuzhou: Fujian Education Press, 1995.
[6] Li Guirong, *Innovative Corporate Culture* [M], Beijing: Economic Management Press, 2002.
[7] Lu Hongfei, *School Culture Construction and Management Research* [M], Shanghai: East China Normal University Press, 2007.
[8] Schein, *Organizational Culture and Leadership* [M], San. Francisco Jossy – Bass Inc., 1992.
[9] Stufflebeam, et al, *Evaluation Model* [M], Trans. Su Jinli, et al, Beijing: Peking University Press, 2007.
[10] Xiang Hongzhuan, *Theories and Practices of School Culture Construction* [M], Hangzhou: Zhejiang University Press, 2010.
[11] Xu Shuye, *Study on School Culture Construction* [M], Guilin: Guangxi Normal University Press, 2008.
[12] Terry Eagleton, *Cultural Conception* [M], Trans. Fang Jie, Nanjing: Nanjing University Press, 2003.
[13] Yu Keping, *Governance and Good Governance* [M], Beijing: Social Sciences Press, 2000.
[14] Zhang Dongjiao, Xu Zhiyong & Zhao Shuxian, *Science of Educational Management* [M], Beijing: Higher Education Press, 2011.

Journal

[1] Chen Li, Analysis on Connotation of Middle and Primary School Development Planning [J], Educational Science Research, 2004 (1).
[2] Feng Wei & Zhao Jianjun, Consideration on Culture construction of Middle and Primary School [J], Journal of Hebei Normal University: Educational Science Edition, 2009 (12).
[3] Fullan, Michael, Erskin-Cullen, Ethne & Watson, Nancy, The Learning Consortium: A School-University Partnership Program: An Introduction [J], School Effectiveness and School Improvement, 1995, 6(3): 187–191.
[4] Gu Mingyuan, On School Culture construction [J], Journal of Southwestern Normal University: Humanities and Social Sciences Edition, 2006 (9).
[5] Ji Ping, Self-examination and Reconstruction of "School Culture", People's Education, 2004 (2).
[6] Jin Zhongming & Lin Chuili, Potential Conflicts of University-Middle and Primary School [J], Shanghai Research on Education, 2006 (6).
[7] Li Jixing, On school motto of middle and primary school [J], Theory and Practice of Education, 2009 (7).
[8] Liu Yunsheng, Considerations on School Culture Planning, Teaching and Management, 2007 (20).
[9] Peng Gang, Self-awareness of School Culture [J], Jiangsu Education, 2004 (21).
[10] Wang Yongchun & Nie Hui, Culture Planning and Its Decisions [J], Consume Guide, 2008 (1).
[11] Wei Changwei, Conflict Management Orientation: Combination of Emergency and Conventionality [J], Theoretical Exploration, 2011 (3).
[12] Xie Yifan, Inheritance and Development: Construction of School Behavioral culture [J], Idea · Theory · Education, 2005 (8).
[13] Yang Xuan, Final Thorough Study on Nine-Year-Sequence Curriculum Reform [J], Guidance of Trends in Education (Taiwan), 2004 (8).
[14] Zhang Zhongguo, Discussion on School Image Design [J], Journal of Chifeng College of Education, 2000 (1).
[15] Zhang Zaiyi & Zhang Minghua, Building a School Culture Brand, Promoting the School Education Quality [J], Journal of Shandong Institute of Education, 2010 (4).
[16] Zhao Lixia, Principles and Contents of School Culture Planning [J], Teaching and Management, 2010 (31).

[17] Zhao Mengcheng, Workplace Learning: Concept, Foundations of Cognition and Teaching Model [J], Comparative Education Review, 2008 (1).

[18] Zheng Rufa, Common Problems in Current Educational Philosophy [J], Teaching and Management, 2009 (1).

[19] Zhong Qiquan, Creation of Principal's Leadership Style and School Culture [J], Educational Reference, 1999 (5).

Dissertation

[1] Yang Ziqiu, Promoting the Research on School Improvement by School-Based Courses Leadership [D], Shanghai: East China Normal University, 2007.

[2] Zhang Feifei, Analysis on Neoliberal Institutionalism Perspective of School Improvement [D], Changchun: East China Normal University, 2008.

[3] Zheng Yulian, Middle-Primary School Principal Training Model Research Based on School Improvement [D], Shanghai: East China Normal University, 2009.

Appendix

List of Schools with School Culture Projects Created

Projects and Time	Schools	
Phase I in Fengtai District, Beijing (2008–2009)	Beijing Fengtai District Donggaodi No. 3 Primary School Beijing Fengtai District Fangguyuan Primary School	Changxindian Central Primary School, Fengtai District, Beijing Experimental Primary School, Fengtai District, Beijing
	Shiliuzhuang Primary School, Fengtai District, Beijing	First Primary School, Fengtai District, Beijing
	Xiluoyuan No. 5 Primary School, Fengtai District, Beijing	Fifth Primary School, Fengtai District, Beijing
	Fanjiacun Primary School, Fengtai District, Beijing	Primary School Attached to Capital Normal University, Fengtai District, Beijing
Phase II in Fengtai District, Beijing (2009–2010)	Dongtieying No. 1 Primary School, Fengtai District, Beijing	Caoqiao Primary School, Fengtai District, Beijing
	Fangchengyuan Primary School, Fengtai District, Beijing	Yungang No. 1 Primary School, Fengtai District, Beijing
	Youanmen No. 1 Primary School, Fengtai District, Beijing	Eighth Primary School, Fengtai District, Beijing
	Xinfadi Primary School, Fengtai District, Beijing	Fengtai Wannianhuacheng Branch of Beijing Primary School
Phase III in Fengtai District, Beijing (2011)	Dongtieying No. 2 Primary School, Fengtai District, Beijing	Yungang No. 2 Primary School, Fengtai District, Beijing
	Donggaodi No. 2 Primary School, Fengtai District, Beijing	Cuilin Primary School, Fengtai District, Beijing
	Jijiamiao Primary School, Fengtai District, Beijing	Chang'an Xincheng Primary School, Fengtai District, Beijing
	Lugou Bridge No. 2 Primary School, Fengtai District, Beijing	Second Primary School, Fengtai District, Beijing
Culture Project of Beijing Teaching Management (2012)	Primary School Affiliated to Capital Normal University	Primary School Affiliated to Chinese Academy of Agricultural Sciences
	Nanhu Zhongyuan Primary School	Fangguyuan Primary School, Fengtai District, Beijing

	No. 1 Primary School Attached to Beijing Xuanwu Normal College	Xiangdong Primary School, Haidian District, Beijing
	Jianhua Experimental Primary School, Beijing	Beijing Fangcaodi International School
	Primary School Affiliated to Beijing No. 1 Normal University	
School Culture Projects Created in Baohe District, Hefei City, Anhui Province (2012)	Hefei Tunxi Road Primary School	Hefei No. 46 Middle School
	Primary School Affiliated to Hefei Normal University	Hefei No. 32 Middle School
	Hefei Sunshine Middle School	
School Culture Projects Created in Baohe District, Hefei City, Anhui Province (2013)	Hefei Wanghu Primary School	No. 3 Primary School Affiliated to Hefei Normal University
	Hefei Experimental School	Hefei No. 28 Middle School
	Hefei No. 29 Middle School	
School Culture Projects Created in Baohe District, Hefei City, Anhui Province (2014)	Hefei No. 48 Middle School	Hefei No. 48 Middle School Binhu Branch Campus
	Hefei Gedadian Primary School	Hefei Weigang Primary School
	Hefei Shuguang Primary School	
Projects in Haidian of Beijing (2008–2010)	Zhongguancun No. 4 Primary School, Haidian District, Beijing	Primary School Affiliated to Capital Normal University
	Beijing No. 2 Experiment Primary School	Peixing Primary School, Haidian District, Beijing
	Shangzhung Zhongxin Primary School, Haidian District, Beijing	Primary School Affiliated to Chinese Academy of Agricultural Sciences
	Cuiwei Primary School, Haidian District, Beijing	Wuyi Primary School in Haidian District, Beijing
	Xiyuhe Primary School, Haidian District, Beijing	
Projects in Shijingshan of Beijing (2011)	Shijingshan District Philharmonic Experimental Primary School, Beijing	Beijing Jingyuan School
	Bajiaobeilu Primary School, Shijingshan District, Beijing	Liyun Experimental School Affiliated to Beijing Normal University

	Gucheng No. 2 Primary School, Shijingshan District, Beijing	Beijing Yangzhuang Middle School
School Culture Projects Created of Beijing Municipal Education Commission (2013)	Beijing HuiWen No. 1 Primary School	Beijing Xicheng Yuxiang Primary School
	Hujialou Central Primary School, Chaoyang District, Beijing	Huajiadi Experimental Primary School, Chaoyang District, Beijing
	Primary School Affiliated to Beijing Medical University, Haidian District, Beijing	Shangdi Experimental Primary School, Haidian District, Beijing
	Fangxingyuan Primary School, Fengtai District, Beijing	Nangong Central Primary School, Fengtai District, Beijing
	Beijing Shijingshan Experimental Primary School	Chengbei Central Primary School, Changping District, Beijing
	Workers' Children Primary School, Mentougou District, Beijing	
School Culture Projects Created by Beijing Municipal Education Commission (2014)	Beijing Dongcheng District Peixin Primary School	Beijing No. 171 Middle School
	Beijing Xuanwu Huimin Primary School	Beijing No. 13 Middle School
	Beijing Haidian District Wanquan Primary School	Beijing Yuying School
	Beijing Chaoyang Jingsong No. 4 Primary School	Beijing Ruifeng School
	Beijing Fengtai District Youanmen No. 1 Primary School	Beijing No. 18 Middle School
	Beijing Shijingshan District Philharmonic Experimental Primary School	Beijing Tongwen Middle School
	Beijing Daxing District No. 9 Primary School	Beijing No. 2 Middle School Yizhuang School
	Beijing Fangshan District Chengguan No. 2 Primary School	Beijing Fangshan District Liangxiang No. 2 Middle School
	Beijing Shunyi District Mapo Central Primary School	Beijing Shunyi District No. 2 Middle School
	Beijing Tongzhou District Liyuan Central Primary School	Beijing No. 2 Middle School Tongzhou Branch Campus
	Beijing Changping District Huilongguan No. 2 Primary School	Beijing Changping No. 1 Middle School

	Beijing Mentougou District Datai Central Primary School	Capital Normal University High School Yongding Branch Campus
	Beijing Miyun Jugezhuang Town Central Primary School	Beijing Miyun Henaizhai Middle School
	Beijing Huairou District Yangsong Middle School	Beijing Huairou District Chawu Railway Middle School
	Beijing Pinggu District No. 2 Primary School	Beijing Pinggu District No. 4 Middle School
	Beijing Yanqing County No. 4 Middle School	Beijing Yanqing County No. 4 Middle School
	Beijing Fangshan District Yanshan Xingcheng Primary School	Beijing Yanshan Qianjin Middle School
Other Project Schools	Beijing Tongzhou District Gongyuan Primary School (2013)	Beijing Haidian Erligou Central Primary School (2013)
	Inner Mongolia Baotou Tuanjiedajie No. 4 Primary School (2013)	Shanxi Xi'an Houzaimen Primary School (2013)
	Xinjiang Urumuqi Foreign Language Primary School (2013)	Beijing Fengtai District No. 8 Middle School (2014)
	Tianjin Ninghe County Banqiao Primary School (2014)	Anhui Wuhu Liminlu Primary School (2014)
	Anhui Hefei Liuanlu Primary School (2014)	

Biographies

Author

Zhang Dongjiao, Professor of educational administration in the Faculty of Education of Beijing Normal University, doctoral supervisor, Ph.D. director of School Culture Research Center of Beijing Normal University, and winner of New Century Excellent Talents in University, is engaged in research fields including school organization culture improvement, principal's leadership skills and competency. She has published over 90 theses on education core journals such as *Educational Research* and *Journal of Beijing Normal University*, as well as published academic monographs about education including *School Culture Improvement Series*, *Theory of Education Communication*, *The Last Totem—Research on Value Orientation in High School Education and School Characteristic Development*, *School Public Relations Management*, *Educational Administration*, and *School Management*. She has independently held over 20 projects at provincial (ministerial) level and won many awards for scientific research at provincial (ministerial) level. Besides, she constructed and applied school culture driving model to guide over 300 elementary and secondary schools to conduct culture practice.

Foreword Writer

GU Mingyuan, Professor, CCP member, was born on 14th October 1929 in Jiangyin county, Jiangsu province. He is now a distinguished professor at Beijing Normal University and a member of the National Education Advisory Committee, the Emeritus President of the Chinese Society of Education, deputy director of the Social Sciences Committee of the Ministry of Education, deputy director of the National Basic Education Curriculum Advisory Committee, deputy director of the Development Research Center of the National Advisory Committee for Education, and director of the Teacher Education Expert Committee of the Ministry of Education.

He once served as assistant director of the Educational Affairs Office at the Affiliated Middle School of Beijing Normal University, principal of the Second Affiliated Middle School of Beijing Normal University, department head of the Department of Education at Beijing Normal University, director of the Institute of Foreign Education Research at Beijing Normal University, vice president of Beijing Normal University, dean of the Graduate School at Beijing Normal University, dean of the College of Educational Administration at Beijing Normal University, convener and member of the 1st, 2nd, 3rd and 4th Education Disciplinary Review Group of the State Degree Committee, director of the National Steering Committee on Ed.M. Degrees, president of the Chinese Society of Education, vice president of China Education Association for International Exchange, and Co-President of the World Council of Comparative Education Societies (WCCES).

Gu Mingyuan is one of the founders of comparative education in China. For 60 years he has adhered to the principle of "Based on China and Broaden Eyes to the Whole World" and is active in education research. He successively finished more than 600 articles and 40 books. Significant works include *LU Xun's Educational Thoughts and Practices* (published in both Chinese and Japanese), *Comparative Education, The Cyclopedia of Education, Education in the USSR After World War II, National Cultural Traditions and Education Modernization, The Collected Edition of Chinese Education, The Dictionary of World Education, My Education Exploration——GU Mingyuan's Educational Collection, Education in China and Abroad: Perspective from a Lifetime in Comparative Education, Cultural Foundations of Chinese Education, Chinese Education Encyclopedia, Educational Thoughts in Luxun's Works, Collection of Professor Gu Mingyuan's Speech* etc. Among them the *Chinese Education Encyclopedia* has been highly praised by Ms Liu Yandong, Vice Prime Minister of China, *"The book achieved depth interpretation of the fundamental theory of education, summed up the historical experience and introduced the latest research results with the very concise and refined language, which played an important role for the promotion of education disciplines construction and enriching socialism educational theory system with Chinese characteristics."*

Professor Gu Mingyuan has not only made great achievements in educational theory but also has had a significant impact in the field of educational practice. He participated in and witnessed almost all of the major education reforms in the 60 years since the founding of the New China, while persistently spreading the love of education throughout the country. He actively participated in international academic exchange and cooperation and achieved great achievements in teaching, research and social service. His outstanding contributions achieved a high degree of recognition in both domestic and foreign academic circles. In 1991 he was named National Prominent Teacher and has enjoyed Government Allowance and Certificate since then. In 1999 he was awarded People's Teacher by Beijing Municipal People's Government. In 2001 he recieved an honorary Doctorate of Education from the Hong Kong Institute of Education, and was named Honorary Professor by the Teachers College of Columbia University in 2008. In 2009 he received honorary doctorate degrees from both the University of Macau and Soka University of Japan. In 2011 he won the Lifetime Achievement Award in the Fourth National Educational Research Awards for Outstanding Achievements, and in 2014 Professor Gu was awarded the Lifetime Achievement Award by the Wu Yuzhang Foundation, the most influential award in the field of social science in China.

Lightning Source UK Ltd.
Milton Keynes UK
UKOW06n2221271015

261491UK00003B/28/P